STATISTICS MANUAL

With Examples Taken From
Ordnance Development

STATISTICS MANUAL

With Examples Taken From
Ordnance Development

By
Edwin L. Crow
Frances A. Davis
Margaret W. Maxfield

Research Department
U. S. Naval Ordnance Test Station

DOVER PUBLICATIONS, INC.
NEW YORK • NEW YORK

U. S. NAVAL ORDNANCE TEST STATION
(An Activity of the Bureau of Ordnance)

D. B. YOUNG, CAPT., USN   WM. B. MCLEAN, PH.D.
Commander   *Technical Director*

Released Within the Department of Defense
as
NAVORD REPORT 3369
NOTS 948

This Dover edition, first published in 1960, is an unabridged, unaltered republication of NAVORD Report 3369—NOTS 948. It is published through special arrangement with the U. S. Naval Ordnance Test Station.

Manufactured in the United States of America

Dover Publications, Inc.
180 Varick Street
New York 14, N. Y.

FOREWORD

This Manual of statistical procedures is intended for use by scientists and engineers. However, especially in the planning stages of tests and experiments, it is advisable to supplement the information by occasional consultations with statisticians.

The proper use of modern statistical methods provides not merely a precise summary of the conclusions that may be drawn from an experiment already performed but also, with a small amount of prior information or by making reasonable assumptions, a reliable prediction of the information that can be gained from a proposed experiment. The latter feature has to some extent already placed the planning of tests and experiments on a substantially more rational basis, and further progress can be expected in the future.

Although the Manual is not intended to be a textbook or treatise on the one hand, or merely a set of tables on the other, it is rendered somewhat self-contained by the inclusion of Chapter 1, Definitions and Distributions, and the Appendix tables and charts. Attention is called particularly to Appendix Table 8, which enables confidence limits for a standard deviation to be obtained immediately, and to Charts II–X, which facilitate the determination of the sample sizes required for various experiments.

The preparation of the Manual was first undertaken at the Naval Ordnance Test Station in 1948, under Task Assignment NOTS-36-Re3d-439-3, as authorized by Bureau of Ordnance letter NP36(Re3d)AAF:bc dated 26 October 1948, which also provided for the basic statistical study of the programs of the Station. Work on the manuscript was not completed at the termination of this task assignment, and the continuation of its preparation from 1950 to 1955 was supported by funds from exploratory and foundational research and general overhead. Although the book was prepared primarily for use at the Station, it may well be of value at other Department of Defense establishments.

The Manual was reviewed for technical accuracy in three sections: Chapters 1, 2, and 3 were reviewed by R. M. McClung, A. W. Nelson, and Nancy L. Seeley; Chapters 4, 7, and 8, by D. E. Hartvigsen, Paul Peach, and R. W. Reynolds; and Chapters 5 and 6, by J. R. Harvey, J. P. Vanderbeck, D. S. Villars, and J. E. Walsh. In addition, the Manual was reviewed as a whole by J. R. Harvey, R. M. McClung, H. A. Meneghelli, Paul Peach, and J. E. Walsh.

Comments and corrections will be appreciated. They should be directed to Commander, U. S. Naval Ordnance Test Station (Code 507), China Lake, California.

D. B. YOUNG, CAPT., USN
Commander

WM. B. McLEAN
Technical Director

INTRODUCTION

The Manual has been prepared for those engineers and scientists who wish to use statistical procedures to improve the efficiency of their experiments. Emphasis is placed on (1) consulting with a statistician while an experiment is in the planning stage, (2) specifying what a decision in a statistical test means by stating the risks involved, and (3) attaching confidence limits to an estimate. The reader who frequently needs to use statistical methods should find a university course helpful. A group of engineers and scientists who desire specific instruction might profitably organize a seminar under the leadership of a statistician.

Users of the Manual may find the following procedure useful:

1. Read Chapter 1 for basic definitions. Since this is a manual rather than a textbook, the explanatory material is brief. References 1.3 and 1.5 are suggested for fuller explanations of points that seem difficult.[1]

2. Study the Table of Contents so that the appropriate part of the book can be found readily when a problem arises. Specific items are listed in the Index.

3. Consult with a statistician while the experiment is being designed. He can indicate the sections in the Manual that describe the proper technique to be used.

4. Read the general explanation of the technique to be used and note the formulas and relevant Appendix tables or charts. Substitute the pertinent data in the general formulas, using a worked example as a guide.

---

[1] Reference numbers refer to the publications listed in the Bibliography section of each chapter. Thus, Ref. 1.5 is the fifth bibliographical reference at the end of Chapter 1. All publications mentioned are available in the Technical Library at the Naval Ordnance Test Station.

5. For fuller treatments, refer to the various texts and papers listed in the Bibliography at the end of each chapter. Derivations and proofs have been omitted in an effort to make the book useful to a reader with only a general mathematical background. (The occasional occurrence of an integral sign should not deter the experimenter who is rusty in mathematics. The sign is merely a convenient symbol for the area under a curve, and the techniques of calculus are not needed.)

The application of statistics in experimentation may be divided into three main steps, as follows:

1. Design of the experiment. Conclusive results from an experiment depend upon an efficient plan. Properly designing the experiment pays dividends in time and money by decreasing the number of computations and increasing the amount of information obtainable. Planning includes the specification of a mathematical statistical model for the phenomena being investigated. It is helpful, but not always possible, to specify the form of the equations; the unknown element then consists of a set of constants called "parameters" (denoted by Greek letters, such as μ and σ).

In nearly all of the methods in the Manual, it is assumed that it is most practical to take a prescribed number of observations. However, if the results of each trial or group of trials can be made available without delay, it is economical to judge the accumulated data at each step by the method of "sequential analysis." Two important types of sequential analysis are given in Sec. 8.4, and many other sequential methods are available in the references.

2. Reduction of data—estimation. By following the appropriate example in the Manual, experimenters will be able to compute characteristics that summarize their data. Such a characteristic of the sample is called a "statistic" and is denoted by an italic letter such as $\bar{x}$ or s. Those statistics which are the best possible estimates (evaluations) of the unknown parameters are of particular interest. Experimenters sometimes use rough methods of data reduction because the best methods are not easily available, but the use of such methods is often equivalent to throwing away a large proportion of the observations. This Manual is intended to make the best methods easily available.

3. Examination of significance. To an extent that depends upon the original experimental design, it will be possible to test hypotheses about the "population" on the basis of the experimental "sample" from that population, or to state the limits of precision in the estimation of parameters. If reasonable care is taken, and if the risks of error are included, these conclusions can be stated rigorously.

ACKNOWLEDGMENT

The authors are indebted to Roland P. Bergeron, Eleanor G. Crow, Edward A. Fay, Robert S. Gardner, Helen C. Morrin, and John W. Odle for detailed study and criticism, calculation of tables, and checking of numerical examples. In addition, the official technical reviewers named in the Foreword suggested many improvements. Acknowledgment is also due to the engineers and scientists who, by the submission and discussion of problems, influenced the selection of topics and examples for the Manual. The many references indicate the obviously great dependence on the extensive open literature dealing with statistical theory and applications.

While revising the manuscript for publication, editors Bernice F. Robinson and June A. Townsend greatly improved the readability, clarity, and accuracy of the presentation. The several typewritten and dittoed versions were prepared by Margaret E. Cox.

The approach to completeness of the Appendix tables and charts was made possible by the generous cooperation of various authors and publishers: R. A. Fisher, F. Yates, and Oliver & Boyd, Ltd., for Tables 3 and 15; Catherine T. Grylls for Table 4; Mrs. Grylls and Maxine Merrington for Tables 5 and 6; E. Lord for Table 13; C. J. Clopper and E. S. Pearson for Charts III and IV; Florence N. David for Chart XI; *Biometrika* for Tables 4–6 and 13, and Charts III, IV, and XI; G. W. Snedecor and Iowa State College Press for Table 7; W. J. Dixon and F. J. Massey, Jr., for Tables 9, 11, and 12; A. H. Bowker for Table 17; McGraw-Hill Book Co., Inc., for Tables 9, 11, 12, and 17; C. Eisenhart and Frieda S. Cohn for Table 10; C. D. Ferris, F. E. Grubbs, and C. L. Weaver for Charts V–VIII; *Annals of Mathematical Statistics* for Table 10 and Charts V–VIII; American Society for Testing Materials for Table 18; Paul Peach and Edwards & Broughton Co. for Table 19; R. S. Burington and A. D. Sprague for Chart I; J. A. Greenwood, Marion M. Sandomire, and *Journal of the American Statistical Association* for Chart IX. In many cases the originals have been abridged or modified, or their notations have been changed, in order to adapt them to the Manual.

Reproduction of copyrighted material is restricted to the Appendixes. Permission to reproduce has been obtained in each instance.

CONTENTS

Chapter 1. DEFINITIONS AND DISTRIBUTIONS

Chapter 2. TESTS AND CONFIDENCE INTERVALS FOR MEANS

Chapter 3. TESTS AND CONFIDENCE INTERVALS FOR STANDARD DEVIATIONS

Chapter 4. TESTS OF DISTRIBUTIONS AS A WHOLE AND ALLIED PROBLEMS

Chapter 5. PLANNING OF EXPERIMENTS AND ANALYSIS OF VARIANCE

Chapter 6. FITTING A FUNCTION OF ONE OR MORE VARIABLES

Chapter 7. QUALITY-CONTROL CHARTS

STATISTICS MANUAL

With Examples Taken From
Ordnance Development

Chapter 1

DEFINITIONS AND DISTRIBUTIONS

1.1. Definitions

1.1.1. Population and Sample

The **population** is the whole class about which conclusions are to be drawn; for instance, a production lot of propellant grains, rather than 100 grains selected as a **sample** from that lot. Frequently it is impossible, or at least impractical, to take measurements on the whole population. In such a case, measurements are taken on a sample drawn from the population, and the findings from the sample are generalized to obtain conclusions about the whole population. For example, if the population is made up of the time to target of every rocket of a certain design, the population may be so large as to be considered infinite. By launching only a sample of the rockets, instead of expending every rocket made, information can still be gained about the time to target of all rockets of the given design.

To permit good generalization to the population, a **random** sample is selected; that is, the sample is chosen in such a manner that every individual in the population has an equal chance of being chosen for the sample. For instance, if there are 1,000 rockets in storage and we want to estimate time to target by firing 50 of them, the 50 should be chosen by lot from among the 1,000, as in a raffle, or by using a table of random numbers (Ref. 1.7) so that each has an equal chance of being one of the 50 fired. In many cases where a true random selection is not feasible, an adequate substitute can be found (Ref. 1.6, Chap. 8). For a test of randomness, see Sec. 4.1 of this Manual.

If separate random samples are drawn, the two samples are **independent** if we select the second sample by lot with no reference to the make-up of the first sample. Thus, in comparing the performance of a rocket of type A with that of a rocket of type B on the basis of a sample of 10 of each type, we should not let the choice of the 10 type-A rockets influence

3

the choice of the 10 type-B rockets in any way. (Sometimes the time at which each observation is made exerts such an influence.) We speak of two variables as being **independent** when fixing the value of one has no effect on the relative frequency with which each value of the other is to be expected. For instance, the horizontal deflection and the vertical deflection of shots at a target are independent if fixing the horizontal deflection has no effect on the probability of obtaining each vertical deflection. Also, in the case of an infinite population, the observations in a random sample are all independent of one another.

When two events are independent, the probability that they both happen can be found by multiplying the probability of the first event by the probability of the second event. ("Probability" is defined in Sec. 1.1.2.) For instance, if the horizontal deflection of a rocket has a probability 0.7 of being less than 10 mils and is independent of the vertical deflection, which has a probability 0.3 of being less than 10 mils, the probability that the rocket will land in a position such that both its horizontal and vertical deflections are less than 10 mils is $0.7 \times 0.3 = 0.21$.

1.1.2. Distribution

To get a good picture of the sample as a whole when the sample is fairly large, it is desirable to construct a **frequency diagram** which will show the number of observations that occur within a given interval.

In Table 1.1, 50 measurements of range of a particular type of rocket are listed; in Table 1.2, the frequency distribution of these measurements

TABLE 1.1. MEASUREMENTS (IN YARDS) OF RANGE OF
A TYPE OF ROCKET

1,472	1,799	1,850	1,251	2,107
1,315	1,372	1,417	1,422	1,668
1,984	1,420	2,057	1,506	1,411
1,900	1,612	1,468	1,651	1,654
1,646	1,546	1,694	1,687	1,503
1,709	1,255	1,776	1,080	1,934
1,780	1,818	1,456	1,866	1,451
1,571	1,357	1,489	1,713	1,240
1,681	1,412	1,618	1,624	1,500
1,453	1,775	1,544	1,370	1,606

TABLE 1.2. FREQUENCY DISTRIBUTION OF
DATA IN TABLE 1.1

Range, yd	Frequency (No. of rounds)
999.5–1,099.5	1
1,099.5–1,199.5	0
1,199.5–1,299.5	3
1,299.5–1,399.5	4
1,399.5–1,499.5	11
1,499.5–1,599.5	6
1,599.5–1,699.5	11
1,699.5–1,799.5	6
1,799.5–1,899.5	3
1,899.5–1,999.5	3
1,999.5–2,099.5	1
2,099.5–2,199.5	1
Total	50

is given. Most of the information is preserved if the data are grouped into intervals of 100 yards, and it is customary to begin the intervals at 999.5, 1,099.5, 1,199.5, etc., in order to show clearly the group in which each measurement belongs. Since the import of the data is more quickly grasped if the frequency distribution is presented in graphical form, a diagram of the data is made, as shown in Fig. 1.1. Such an arrangement can easily be made with a typewriter.

It is often of more interest to know the **relative frequency** or **percentage frequency** (100 times the relative frequency) of observations falling in a given interval than the actual number. In Fig. 1.2, which is a **histogram** or **bar diagram**, the percentage frequencies as well as the actual frequencies are given. (If there were more sample members and shorter intervals, the outline of the histogram would probably show a smoother curve.)

In dealing with a population, rather than a sample, we use the theoretical equivalent of the relative frequency; that is, the **probability** of observations falling in a given interval or, more generally, the probability of any event out of a given set of events. Since the probability of an event E, denoted by $P(E)$, is essentially defined as the relative frequency

RANGE IN YD	FREQUENCY			
	0	5	10	15
1,000				
	X			
1,100				
1,200				
	XXX			
1,300				
	XXXX			
1,400				
	XXXXXXXXXX			
1,500				
	XXXXX			
1,600				
	XXXXXXXXXX			
1,700				
	XXXXX			
1,800				
	XXX			
1,900				
	XXX			
2,000				
	X			
2,100				
	X			
2,200				

FIG. 1.1. A Graphical Presentation of a Frequency Distribution. (Precise intervals begin at 999.5, 1,099.5, etc.)

of occurrence of E in a long series of trials, any probability lies in the range from 0 to 1. Probabilities can be operated on according to two basic laws. For any two events, denoted by A and B:

(1) $P(A \text{ and } B) = P(A) \times P(B \text{ if } A \text{ has occurred})$
$= P(B) \times P(A \text{ if } B \text{ has occurred})$

(1a)   If A and B are *independent* events
$$P(A \text{ and } B) = P(A) \times P(B)$$
(Examples at end of Sec. 1.1.1 and 1.2.1.)

(2) $P(A \text{ or } B \text{ or both}) = P(A) + P(B) - P(A \text{ and } B)$

(2a)   If A and B are *mutually exclusive* events
$$P(A \text{ or } B \text{ or both}) = P(A \text{ or } B) = P(A) + P(B)$$

These can be extended to any number of events (Ref. 1.2).

As an analog of the sample frequency distribution, we introduce the **population frequency distribution** or **probability distribution** of a variable x, often called merely the **distribution** of x. (In the example given at the beginning of this section, x would be the range of the rocket.) The distribution of x shows the relative frequencies with which all possible values of x occur. A population distribution can be pictured in terms of a curve $f(x)$, which is well approximated by a relative-frequency histogram, made with rectangles having unit width, for a large sample from the population.

To see the precise meaning of the curve $f(x)$, let ΔF be the probability that an observation will fall in the interval of length Δx extending from

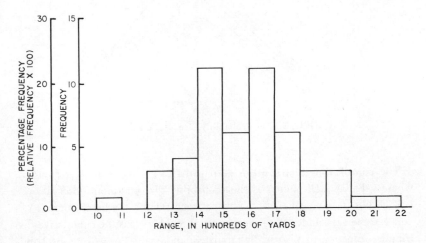

FIG. 1.2. A Histogram.

x to $x + \Delta x$. Then $\Delta F / \Delta x$ can by physical analogy be called the average probability density in the interval. If we let Δx approach zero, the average density can be considered as approaching a limiting value $f(x)$, which we call the probability density at x. This is represented in symbols by the equation

$$\lim_{\Delta x \to 0} \frac{\Delta F}{\Delta x} = \frac{dF}{dx} = f(x)$$

We therefore call $f(x)$ the **probability density function** of x. It is also called the **distribution function** of x or, loosely, the **distribution** of x. When $f(x)$ is drawn as a graph (Fig. 1.3), the probability that a random observation x will fall in any interval a to b is the area under the curve $f(x)$ from a to b. This area is expressed as an integral

$$P(a < x \leqslant b) = \int_{a}^{b} f(x) \, dx$$

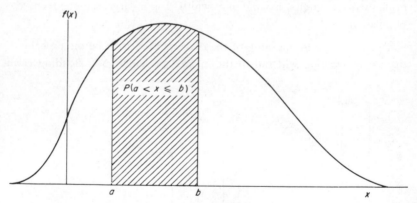

FIG. 1.3. A Probability Density Function.

The **cumulative distribution function** $F(x)$ (Fig. 1.4) is a function whose value at each point c is the probability that a random observation x will be less than or equal to c. Thus, at each point c the height of $F(c)$ is equal to the area under $f(x)$ from $-\infty$ through c. It is the cumulative distribution function that is usually tabulated, and enables us to get a probability over any interval simply by subtracting one area from another:

$$P(a < x \leqslant b) = \int_{-\infty}^{b} f(x)\,dx - \int_{-\infty}^{a} f(x)\,dx = F(b) - F(a)$$

Since a random observation x is sure to have some value, the total area under the curve $f(x)$ is unity:

$$\int_{-\infty}^{\infty} f(x)\,dx = 1$$

Correspondingly, the cumulative distribution function approaches 1 as x gets large: $F(+\infty) = 1$. Likewise $F(-\infty) = 0$. Also, the probability density function $f(3)$, for example, is not itself the probability of obtaining a random reading of 3, since the probability of obtaining a particular mathematical number for a random observation is zero; that is, for any number a, the area at a is

$$\int_{a}^{a} f(x)\,dx = 0$$

This discussion of probability distributions has considered just one of two main types, in which the variable may take all values in an interval and

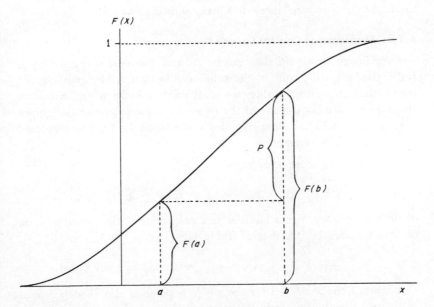

FIG. 1.4. A Cumulative Distribution Function.

is represented by a smooth curve. In the other main type, the variable takes on only discrete values; e.g., the non-negative integers 0, 1, 2, $\cdots$. The two types of distributions are called **continuous** and **discrete** distributions, respectively. An example of the discrete type is obtained by making repeated tests of 10 fuzes in many different lots. In each test, x fuzes are found defective, where x may take on any one of the 11 values 0, 1, 2, $\cdots$, 10, but no others (in particular, no fractional values). If a sample of 10 fuzes is drawn from each of 50 lots and the 50 samples are tested, a histogram may be constructed (Fig. 1.5) just as for the rocket ranges above, each rectangle being of unit width and centered at an integer. However, the abutting rectangles are drawn merely for visual interest, and it should be remembered that each frequency or relative frequency is concentrated at a point.

If infinitely many lots could be tested, we would obtain the theoretical model (the probability distribution), but this distribution would still be represented graphically by rectangles rather than by a smooth curve. See Sec. 1.2.2 on the binomial distribution for the most important example of a discrete probability distribution; this distribution can be used as a model for the fuze-testing problem above and many similar problems.

1.1.3. Measures of Central Location

As indicated above, the data can be reduced considerably by grouping, and a good over-all picture of the values can be obtained by graphing. To reduce the sample data further, we shall want to know where the values are centered. We can use one of the measures in the following paragraphs (Sec. 1.1.3a–1.1.3d) to summarize that information. Of the four measures, the first is the most common.

a. Mean. The mean or average

$$\bar{x} = \frac{1}{n}(x_1 + x_2 + \cdots + x_n) = \frac{1}{n}\sum_{i=1}^{n} x_i$$

of the n sample readings is familiar as a description of the central location of school grades. For the data of Table 1.1

$$\bar{x} = \frac{1}{50}(1{,}472 + 1{,}315 + \cdots + 1{,}606) = 1{,}589.8 \text{ yards}$$

A mean μ can be defined for the whole population also, though it must be defined slightly differently if the population is infinite. Physically, the

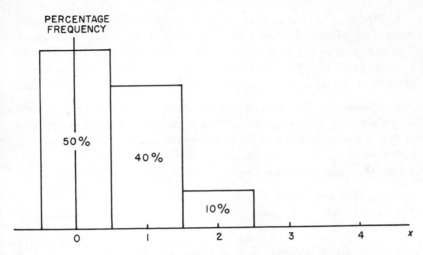

FIG. 1.5. A Discrete Probability Distribution.

mean μ is the x-coordinate of the center of gravity of the area under the probability density curve for the population. The mean $\bar{x}$ of the sample readings is generally the best estimate of the mean μ of the population from which the readings were drawn.

b. Median. The median is the halfway point in the readings when they have been arranged in order of size (the middle reading of an odd number of readings, or the average of the middle two for an even number). For the data of Table 1.1, the median is 1,588.5 yards, halfway between 1,571 and 1,606.

The population median is the fiftieth percentile of the population; that is, the number $\tilde{m}$ such that the area under the distribution curve to the left of $\tilde{m}$ is 0.5.

The population median and mean may not be equal. In the case of the normal distribution (Sec. 1.2.1), where the two are equal, the sample *median* may be used to estimate the population *mean;* however, the median is not so good an estimate as is the sample mean, in the sense that it does not draw the maximum amount of information from the sample. The median is less affected by a few wild readings than is the mean.

c. Mode. A mode is a peak value of the frequency distribution. A single mode of high frequency gives a rough but quick measure of central location. The idea of mode is not useful, however, if the distribution has more than one high point, as in Fig. 1.2, or does not show a single, well-defined peak. More than one high point may indicate the need for regrouping the data.

d. Midrange. The midrange is the average of the smallest and the largest readings. This is a good measure of the central location for samples of five or fewer, though it is not so good as the mean.

1.1.4. Measures of Dispersion

In reducing sample data, it is also necessary to know how the values are distributed about their center. There are several common measures of dispersion:

a. Standard Deviation. The standard deviation s of a sample of n observations $x_1, x_2, \cdots, x_n$ is

$$s = \left[\frac{1}{n-1} \sum_{i=1}^{n} (x_i - \bar{x})^2 \right]^{\frac{1}{2}}$$

The **variance** s^2 of the sample values is the square of s; that is, essentially the average of the squares of distances from the mean—aside from a factor $n/(n-1)$.

The population parameters corresponding to s and s^2 are called the population standard deviation and variance, respectively, and are denoted by σ and σ^2, following the convention of using Greek letters for population parameters.

The reason for defining the sample variance s^2 with $n-1$ rather than n in the denominator is that the sample variance is of most importance as an estimate of the population variance σ^2 and, when defined as above, it is an **unbiased** estimate of σ^2. "Unbiased" indicates that the values of s^2 from all the hypothetically possible samples from the same population will average to σ^2. The sample variance s^2 is the best estimate of σ^2 in the case of a normal population. (As an estimate of σ, s is slightly biased. See Ref. 1.3, p. 72.)

The computing of s^2 is simplified by using the algebraically equivalent form

$$s^2 = \frac{1}{n(n-1)} \left[n \sum_{i=1}^{n} x_i^2 - \left(\sum_{i=1}^{n} x_i \right)^2 \right]$$

For the data of Table 1.1

$$s^2 = \frac{50(1,472^2 + 1,315^2 + \cdots + 1,606^2) - (79,490)^2}{(50)(49)} = 48,683$$

$$s = 220.6$$

b. Range. The range w is the difference between the smallest and the largest readings in the sample. In the case of a normal distribution (Sec. 1.2.1), the range multiplied by the appropriate factor from Appendix Table 12 gives a good measure of the population σ for a small sample ($n \leqslant 10$), though not so good as s (Ref. 1.3, p. 239). For the data of Table 1.1, the range is $2,107 - 1,080 = 1,027$. Using the factor 0.222 from Table 12, we get an estimate $0.222 \times 1,027 \cong 228$ for σ.

c. Mean Deviation. The mean deviation is the average of the distances from the mean of the n sample members (all distances being taken as positive). The mean deviation is also often defined as the average of distances from the *median* rather than from the mean. For estimating σ in a normal population, the mean deviation is not so efficient as s, but can be multiplied by a factor (Ref. 1.3, p. 240) to give a fairly good estimate for sample sizes $n \leqslant 10$.

d. Various Percentile Estimates. Estimates of the population σ based on percentiles of the sample can be made, but they are not so good as the sample standard deviation. For instance, $(P_{93} - P_{07})(0.3388)$ can be used. To find P_{93}, arrange the values in order of size and count off 93% of them. The value that appears nearest that point is used for P_{93}. Other percentile estimates using more than two percentiles (and consequently deriving more information from the sample) can be found in Ref. 1.3, p. 231. For the data of Table 1.1, we find $(P_{93} - P_{07})(0.3388) = 230$, as compared with $s = 220.6$.

e. Probable Error. The probable error (PE) is a deviation from the population mean μ such that 50% of the observations may be expected to lie between $\mu - PE$ and $\mu + PE$. *For the normal distribution*, Appendix Table 2 shows that the probable error equals 0.674σ. The probable error can be converted to standard deviation by multiplying by 1.4826. The more modern standard deviation is in wider use as a measure of dispersion than is the probable error, because the standard deviation arises naturally in the derivations that underlie the statistical tests in common use.

1.1.5. Coding

Computation of the measures defined in Sec. 1.1.3 and 1.1.4 can often be simplified by **coding** the data; that is, by replacing the readings themselves by coded numbers (usually shorter) at the beginning of computation and decoding at the end. The following methods may be used: (1) subtracting a constant, (2) multiplying by a constant, or (3) subtracting a constant and multiplying by a constant. For instance, in computing the mean of the readings in Table 1.1, subtracting 1,000 from each reading can simplify the procedure. Thus, instead of

$$\bar{x} = \frac{1}{50} (1,472 + 1,315 + 1,984 + \cdots + 1,606) = 1,589.8$$

the equation would read:

$$\bar{x} = 1,000 + \frac{1}{50} (472 + 315 + 984 + \cdots + 606) = 1,589.8$$

Another example of coding is given in Sec. 2.2.2.

1.1.6. Sampling Distribution

Suppose we consider a population of 300 individuals. Taking all possible samples of 10 (there would be $300!/(10! \times 290!)$, or more than 10^{18} of them) and recording the sample $\bar{x}$ of each would give a population of $\bar{x}$'s, one for each sample of 10. That population would have a frequency distribution, called the **sampling distribution**, of $\bar{x}$ for samples of size 10. Similarly, we have sampling distributions for s^2, or for any other sample statistic. The sampling distribution of $\bar{x}$ has a mean and a standard deviation, which can be denoted by $\mu_{\bar{x}}$ and $\sigma_{\bar{x}}$. The standard deviation of a sampling distribution is usually called the **standard error** of the statistic. For sampling from any infinite population of x's, it can be shown, for example, that the mean $\mu_{\bar{x}}$ of the distribution of $\bar{x}$ is equal to the mean μ of the population of x's and that $\sigma_{\bar{x}}$, the standard error of $\bar{x}$, is equal to $\sigma/\sqrt{n}$, where σ is the standard deviation of the population of x's and n is the number of observations in each sample. Thus $\bar{x}$ will tend to approximate μ more closely, the larger the sample size.

1.1.7. Test of a Statistical Hypothesis

We often draw and analyze a sample for the purpose of testing an initial hypothesis, called the **null hypothesis**, about the population. For example, we might state as a null hypothesis that the variance σ^2 equals 15. This might represent a coded variance of 0.0015 in² on a molded part. It can be shown that the fraction $(n - 1)s^2/\sigma^2$ has a certain sampling distribution, which can be found in tables. For example, if σ^2 is indeed 15 and the population has a certain type of distribution (the normal), then the statistic $(n - 1)s^2/15$ computed from samples of size 11 has the sampling distribution shown in Fig. 1.6. The tabled distribution so graphed shows that the probability that the statistic $10s^2/15$ falls below 3.25 or above 20.5 (that is, the area under those portions of the probability density curve) is only 5% if σ^2 is actually 15.

To perform the test, we compute $10s^2/15$ from the sample. If it falls outside the interval 3.25 to 20.5, we reject the null hypothesis that σ^2 is 15

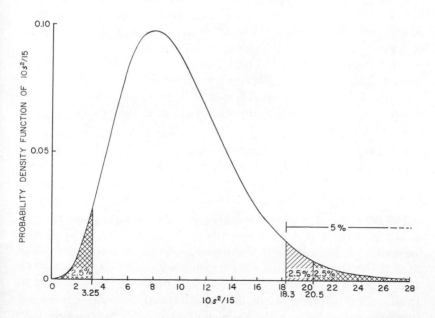

FIG. 1.6. Frequency Distribution of $(n - 1)\, s^2/\sigma^2$ (for $n = 11$ and $\sigma^2 = 15$).

on the ground that if σ^2 were 15, such a large or small value of $10s^2/15$ would be quite improbable.

In this example we have rejected the hypothesis whether $10s^2/15$ was too large *or* too small, by using what is called an **equal-tails test**, with 2.5% at each end of the area under the distribution curve included in the rejection region. If we were interested only in whether the variance σ^2 was not *greater* than 15, we would use the **one-sided test** with 5% rejection region $10s^2/15 > 18.3$.

If we do not reject the null hypothesis, we do not then categorically conclude that the hypothesis is true (see the last paragraph of Sec. 1.1.8). We merely recognize that the sample is compatible with the kind of population described in the null hypothesis; that is, that the statistic computed cannot be considered extreme if the null hypothesis holds.

1.1.8. Level of Significance α and Types of Error

In the example given above, we reject the null hypothesis if the observed $10s^2/15$ falls in a region of extreme values to which the null hypothesis assigns probability only 0.05. Here, 5% is called the **level of significance** of the test and is, in general, denoted by α. There is a 5% risk of error, for even if the null hypothesis does hold, there is a 5% probability that we will reject it. This type of error, the rejection of the null hypothesis when it is true, is called a **Type I error.** The risk of Type I error is α, the level of significance. This level is arbitrary, but should be chosen when the experiment is designed.

Sometimes when it is important to guard against Type I errors, as in testing an expensive product, a level α of 1% or 0.1% is used. The level of 5% is used throughout this Manual. (In some texts, results that are significant at the 5% level are marked with a star.)

The failure to reject a false hypothesis is called a **Type II error.** The risk of a Type II error is denoted by β. In the example above, suppose that the variance is not equal to 15, but has some alternative value, say 25. It is possible that the statistic $10s^2/15$ will fall between 3.25 and 20.5, leading us to accept (fail to reject) the hypothesis that σ^2 is 15, though it is actually 25. The probability of making a Type II error depends on the alternative value; for example, we would be less likely to accept the null hypothesis when σ^2 equals 25 than when σ^2 equals 16.

1.1.9. Operating-Characteristic Curve

An operating-characteristic (OC) curve shows the probability of a Type II error for different alternatives. Examples appear in Appendix Charts V–VIII. These curves are useful in determining the sample size for an experiment. First, we choose the level of significance α at which a test is to be conducted (i.e., the risk that we can take of rejecting the null hypothesis if it is true). Then we decide on an alternative that we wish to detect or guard against and the risk β that we are willing to take of making a Type II error. (It is at this step that we key *statistical* significance to *practical* significance in the problem at hand.) The OC curves will show what sample size will satisfy the two conditions. The confidence interval, explained below, is also useful in determining sample size. Details of these two methods of determining sample size are given, with particular problems, in following chapters of this Manual.

Most of us have some intuition about permissible risks in ordinary problems, but some experimenters may object that it is about as difficult to specify the risks α and β as to specify a sample size, or more difficult if a budget has already been fixed. Even if the sample size has already been specified, it is wise to determine the consequent risks from the appropriate OC curve.

An alternative, but more difficult, method of determining sample size is to choose it so that the total cost is at a minimum. If we denote the null hypothesis under test by H_0 and the alternative we wish to detect by H_1, and we make final acceptance of one or the other on the basis of the experiment, then the total cost is given by the equation

Total cost $=$ (risk of accepting H_0 if H_1 is true)
$\times$ (cost of accepting H_0 if H_1 is true)
$+$ (risk of accepting H_1 if H_0 is true)
$\times$ (cost of accepting H_1 if H_0 is true)
$+$ (cost of experiment)

As the sample size increases, the first two terms decrease (because the risks decrease), and the last term increases. Thus, the curve of total cost against sample size may be expected to decrease at first but to increase ultimately. The sample size giving the minimum point is the appropriate choice. The risks can be evaluated by the theory of probability, but the costs must be specified by the user of the experimental results. This specification may be difficult; consider, for example, the danger of accepting the poorer of two weapons for general use for an unknown number of years to come. However, the approach may be of value in many problems.

1.1.10. Confidence Intervals

In estimating a population measure, say μ, by a sample measure, say $\bar{x}$, we need not be content with recording that μ has a value somewhere near $\bar{x}$. Actually, we can make use of the sampling distribution of $\bar{x}$ to construct an interval about $\bar{x}$ and, with a specified confidence, state that μ lies in the interval. For a 95% **confidence interval**, the statement will be correct 95% of the time, in the sense that if we drew many samples and computed $\bar{x}$ and the corresponding confidence interval for each of them, in about 95 cases out of 100 the confidence intervals would contain μ.

The end points of a confidence interval are called **confidence limits**, and the relative frequency with which the interval will contain μ (in the sense cited above) is called the **confidence coefficient**. (The confidence coefficient is frequently chosen as 0.95 or 0.99.) By specifying the length of the confidence interval, we can often determine sample size.

If the standard error of a statistic is known or can be estimated, an approximate 95% confidence interval can be constructed easily by using as confidence limits the statistic plus or minus twice its standard error. For example, such confidence limits for the mean are $\bar{x} \pm 2\sigma_{\bar{x}} = \bar{x} \pm 2\sigma/\sqrt{n}$, where σ is the standard deviation of an individual observation. If σ is known, we can use the length of the confidence interval, $4\sigma/\sqrt{n}$, to determine the sample size n. Thus, if σ is 10 and we want to hold μ to an interval of length 5, we have $4(10)/\sqrt{n} = 5$, or $n = 64$.

1.1.11. Tolerance Intervals

Confidence limits enclose *some population parameter* (μ, for instance) with a given confidence. **Tolerance intervals** are constructed from experimental data so as to enclose *P% or more of the population* with the given confidence $1 - \alpha$. That is, if tolerance intervals are constructed for many samples according to the specified rules, then in $100(1 - \alpha)\%$ of the cases they will enclose at least $P\%$ of the population. (See Sec. 4.9.) The percentage P may be called the **population coverage**.

The tolerance intervals discussed here are computed from the sample to show where most of the population can be expected to lie. They are, then, different from the tolerances given in specifications which show where all acceptable items *must* lie.

1.1.12. Degrees of Freedom

In the language of physics, a point that can move freely in three-dimensional space has three degrees of freedom. Three variable coordinates, x, y, and z, can take on different values independently. If we constrain the point to move in a plane, it has just two degrees of freedom, this fact being shown by a dependency relation among x, y, and z: $ax + by + cz = d$.

A similar concept is used in statistical language. The sum of n squares of deviations from the sample mean

$$\sum_{i=1}^{n} (x_i - \bar{x})^2 = (x_1 - \bar{x})^2 + \cdots + (x_n - \bar{x})^2$$

where

$$\bar{x} = \frac{1}{n} \sum_{i=1}^{n} x_i$$

is said to have just $n - 1$ degrees of freedom, for if $\bar{x}$ is fixed, only $n - 1$ of the x's can be chosen independently and the nth is then determined. The sampling distributions of some statistics depend on the number of degrees of freedom. The most common examples of such statistics are t, or Student's t (Chapter 2), χ^2 (Chapters 3 and 4), and F (Chapters 3 and 5). Their distributions are tabulated in the Appendix.

1.2. Particular Distributions

1.2.1. Normal Distribution

The normal distribution, which has the probability density function

$$f(x) = \frac{1}{\sqrt{2\pi}\,\sigma} e^{-(x-\mu)^2/2\sigma^2}$$

with mean μ and standard deviation σ, is of particular interest in both applied and theoretical statistics. It occurs frequently in practical problems and is easy to use because its properties have been thoroughly investigated.

The normal distribution is illustrated in Fig. 1.7 in terms of the standard-ized variable $z = (x - \mu)/\sigma$. The transformation or change of variable $z = (x - \mu)/\sigma$ is in common use, for it permits us to use one table for all normal distributions, irrespective of their different means and variances. The distribution of z is normal with zero mean and unit standard deviation. The normal curve is symmetric, with its one mode (or peak) and its median at the mean μ.

If each of several independent variables x_i has a normal distribution with mean μ_i and variance σ_i^2, then the sum $y = \Sigma c_i x_i$, where the c_i are constants, has a normal distribution with mean $\Sigma c_i \mu_i$ and variance $\Sigma c_i^2 \sigma_i^2$.

The cumulative distribution function of the normal distribution is

$$F(x) = \frac{1}{\sqrt{2\pi}\,\sigma} \int_{-\infty}^{x} e^{-(x-\mu)^2/2\sigma^2} dx$$

A typical curve of this type is given in Fig. 1.8. Note that the height of the curve $F(x)$ at a point c equals the area under the curve $f(x)$ to the left of c.

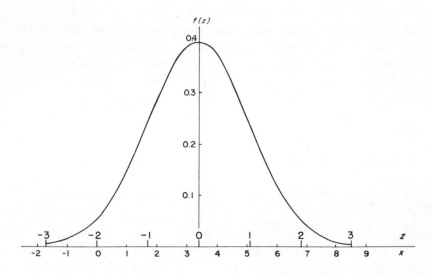

FIG. 1.7. Probability Density Function of the Normal Distribution.

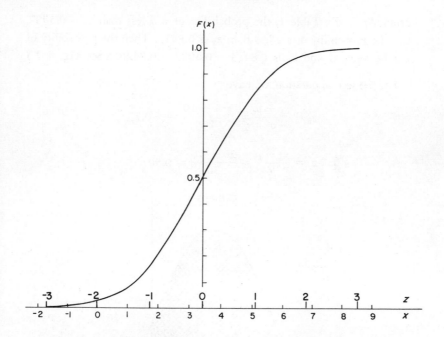

FIG. 1.8. Cumulative Distribution Function of the Normal Distribution.

Values of the cumulative distribution function corresponding to values of z are given in Appendix Table 1 along with a sketch showing the area under the normal curve as it appears in a particular case. Values of z corresponding to an area F are given in Appendix Table 2. The following examples illustrate the use of these tables.

Example. If x has a normal distribution, mean 10, and standard deviation 20, what is the probability of a random observation falling between 15 and 30? Between -5 and 20?

For the first question, we find

$$z_1 = \frac{x_1 - \mu}{\sigma} = \frac{15 - 10}{20} = 0.25$$

$$z_2 = \frac{x_2 - \mu}{\sigma} = \frac{30 - 10}{20} = 1.0$$

From Appendix Table 1, the probability of a z less than z_1 is 0.5987, and the probability of a z less than z_2 is 0.8413. Then the probability of a z between z_1 and z_2 is $0.8413 - 0.5987 = 0.2426$. (See Fig. 1.9.)

For the second question, we have

$$z_1 = \frac{x_1 - \mu}{\sigma} = \frac{-5 - 10}{20} = -0.75$$

$$z_2 = \frac{x_2 - \mu}{\sigma} = \frac{20 - 10}{20} = 0.50$$

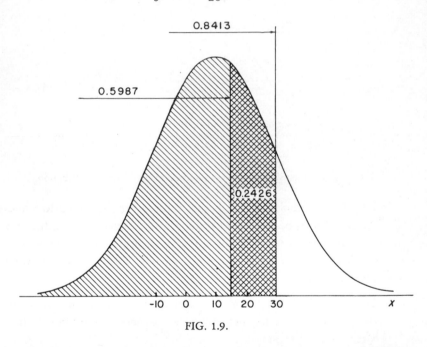

FIG. 1.9.

From symmetry, the probability of a z *less than* 0.75 is the same as the probability of a z *greater than* -0.75. From Appendix Table 1, the probability of a z less than 0.75 is 0.7734; hence the probability of a z less than -0.75 is $1 - 0.7734 = 0.2266$. The probability of a z less than z_2 is 0.6915. The probability of a z between z_1 and z_2 is then $0.6915 - 0.2266 = 0.4649$. (See Fig. 1.10.)

Example. What is the probability that a random normal number will lie within 0.7σ units of the mean?

We compute
$$z = \frac{x - \mu}{\sigma} = \frac{(\mu + 0.7\sigma) - \mu}{\sigma} = 0.7$$

From Appendix Table 1, the probability of a z less than 0.7 is 0.7580. The probability of a z less than zero is 0.5000. Subtracting, the probability

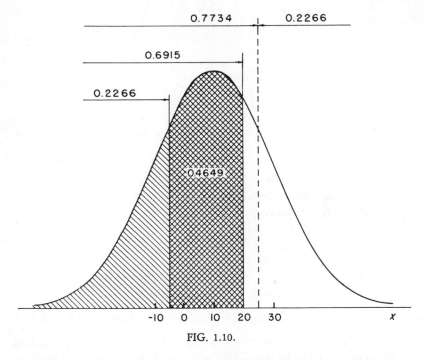

FIG. 1.10.

of a z between zero and 0.7 is $0.7580 - 0.5000 = 0.2580$. From symmetry, the probability of a reading within 0.7 to the left of the mean is also 0.2580. The total probability, then, is $2(0.2580) = 0.5160$. (See Fig. 1.11.)

Example. If x has a normal distribution with standard deviation 10, find a balanced interval around the mean so that the probability is 0.95 that a random reading will fall in the interval.

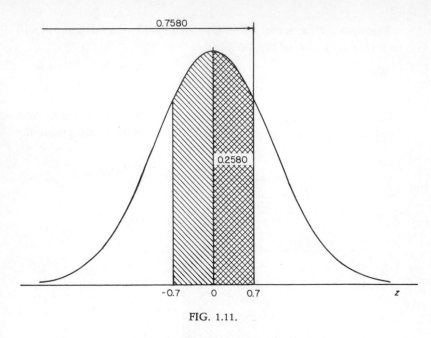

FIG. 1.11.

Since an area of $(1 - 0.95)/2 = 0.025$ must lie to the left of the interval, we read from Appendix Table 2 that $z = -1.960$. From

$$z = \frac{x - \mu}{\sigma}$$

we have

$$x = \mu + z\sigma = \mu + (-1.960)(10) = \mu - 19.60$$

The desired interval is from $\mu - 19.60$ to $\mu + 19.60$. (See Fig. 1.12.)

From Appendix Table 2, we see that 50% of a normal population falls within 0.674σ of the mean. From Appendix Table 1, 68.3% falls within 1σ of the mean; 95.4% within 2σ; and 99.7% within 3σ.

Some of the uses of the normal distribution function are listed below. It can be used:

1. As a curve to "fit" a set of data which appear to be approximately normally distributed when the sample is fairly large (> 50). (See Chapter 4 for tests of normality.) The cumulative normal distribution

is "fitted" by substituting the sample mean and standard deviation for the corresponding population values in the equation for $F(x)$.

2. As an approximating distribution for certain sample statistics, such as the mean, when the sample size is fairly large (> 30 for the mean).

3. As an approximation to other distributions such as the binomial (to be discussed subsequently) when the samples are fairly large and certain other criteria are satisfied. (See Ref. 1.11, pp. 152–59.)

4. As the basic distribution in most practical applications, for tests of significance, quality-control work, design of experiments, and variables sampling; that is, the variable x under consideration is assumed to be normally distributed. (See Chapter 4 for tests of normality.)

One of the applications of considerable interest on this Station is determining the probability of a hit on a target when the shots have a normal distribution. The technique to be applied depends on the shape and position of the target. The following examples are typical.

Example (One Dimension). A plane flies low above a highway and drops bombs. Call the perpendicular distance from the highway center to each bomb-impact point its x-distance (positive to one side of the road, negative to the other), and assume that the variable x has a normal

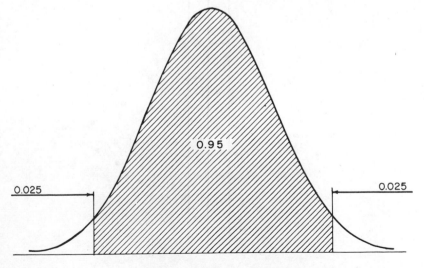

FIG. 1.12.

distribution with mean zero and standard deviation 20 yards. What proportion of the bombs dropped can be expected to fall within 6 yards of the center of the highway? (See Fig. 1.13.)

We have $z_1 = (-6 - 0)/20 = -0.3$, and $z_2 = (6 - 0)/20 = 0.3$. From Appendix Table 1, $F(0.3) = 0.6179$. Then the expected proportion of successful shots (within 6 yards of the highway center) is $2(0.6179 - 0.5) = 0.2358$.

Example (Two Dimensions, Rectangular Target). Suppose shots aimed at a rectangular target are subject to random influences such that

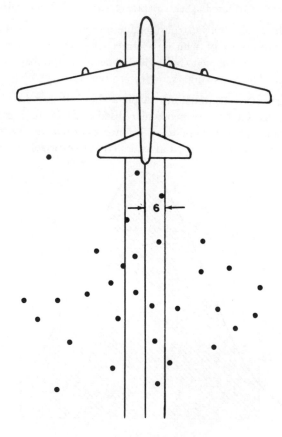

FIG. 1.13.

the horizontal deviations x and the vertical deviations y of the shots are independent and normally distributed. (See Fig. 1.14.) What proportion of shots can be expected to hit the target?

Since the horizontal coordinate and the vertical coordinate are independent, the probability that a particular shot will fall in the target area is the probability that its horizontal coordinate will fall in the interval (a, b) cut off by the target, denoted by $P(a < x < b)$, multiplied by the probability that its vertical coordinate will fall in the interval (c, d) cut off by the target, denoted by $P(c < y < d)$. Thus the expected proportion is $P(a < x < b) \cdot P(c < y < d)$. Each of these probabilities can be found by use of Appendix Table 1. For instance, if the mean values of x and y are both zero and $a = -0.6745\sigma_x$, $b = 0.6745\sigma_x$, $c = -0.6745\sigma_y$, and $d = 0.6745\sigma_y$, then $P(a < x < b) = 0.5000$ and $P(c < y < d) = 0.5000$. Hence, the probability of a shot lying in the corresponding rectangle is 0.2500.

Example (Two Dimensions, Circular Target). Suppose shots aimed at a circular target are subject to random influences such that the hori-

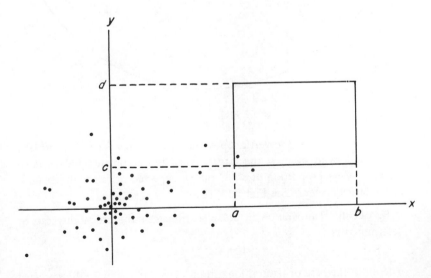

FIG. 1.14.

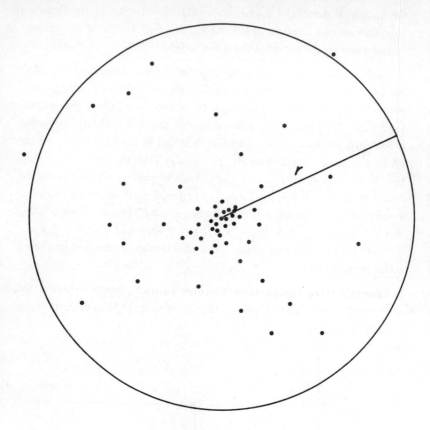

FIG. 1.15.

zontal positions and the vertical positions are independent and normally distributed, with mean at the target center and common standard deviation σ. What proportion of shots can be expected to fall within r units of the target center? (See Fig. 1.15.)

By calculus (integration in polar coordinates) the probability can be determined as

$$1 - e^{-r^2/2\sigma^2}$$

where e is the base of natural logarithms. Thus, if r is twice the standard deviation, we have

$$1 - e^{-(2\sigma)^2/2\sigma^2} = 1 - e^{-2} = 1 - 0.135 = 0.865$$

as the proportion expected to hit the target in this case. The radius of the circle centered at the mean which contains 50% of the shots (called the **circular probable error**) is $\sigma\sqrt{2\log_e 2} = 1.1774\sigma$. The root-mean-square radial error—that is, the square root of the average of the squared radial distances from the mean point—is $\sigma\sqrt{2} = 1.4142\sigma$.

Example (Two Dimensions, Any Target). For other target shapes and positions a chart technique is available. Again we assume that the horizontal and vertical deviations have independent normal distributions. The target is drawn to scale on tracing paper and held over Appendix Chart I.[1] (In practice, it is preferable to use a suitable copy of Appendix Chart I mounted on cardboard.) A count of the number of rectangles covered by the target gives the probability of a hit, each rectangle representing a probability of 0.001. The scales for drawing the target, possibly different for each of the two perpendicular directions, are fixed by the condition that the standard deviation of the shots in each direction must correspond to the length marked on the chart. Since the chart shows only one quadrant, it must be rotated through four settings to give a complete count.

Suppose that points of impact of a type of rocket have a two-dimensional normal distribution about the target point with standard deviation of 200 yards in range and 60 yards in lateral deflection. What is the probability that a rocket will land within a circle of 100-yard radius about the target point?

In order to use Appendix Chart I to obtain the probability, we must first draw the target to scale, so that the lengths marked "1 standard deviation" on the chart will correspond in the range direction to 200 yards and in the lateral direction to 60 yards. The circle becomes an ellipse, one quadrant of which (superimposed on Chart I) is shown in Fig. 1.16. A count of the chart rectangles in this quadrant covered by the ellipse gives 73.3. Multiplying by 4 to allow for the other (symmetric) quadrants, we have 293.2. Since each rectangle corresponds to a probability of 0.001, the probability of a hit within 100 yards of the target is about 0.293.

---

[1] The rectangular chart was introduced in the Bureau of Ordnance during World War II by R. S. Burington, E. L. Crow, and A. D. Sprague.

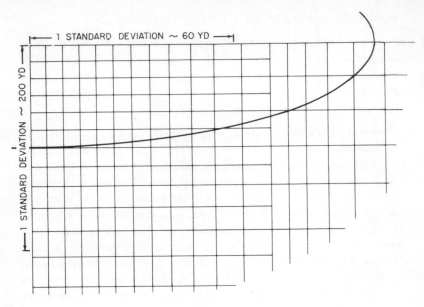

FIG. 1.16.

1.2.2. Binomial Distribution

Suppose each sample member is to be judged on some yes-or-no basis, such as a hit or a miss, a good item or a defective item, a head or a tail. Suppose, further, that a random item has probability p of being a hit and probability q of being a miss. Since it must be one or the other, we have $p + q = 1$. The binomial distribution is defined by

$$P_n(x) = C_x^n p^x q^{n-x} \qquad (x = 0, 1, 2, \cdots, n)$$

where

$P_n(x) =$ probability of getting *exactly* x hits in n trials
$\quad C_x^n =$ number of combinations of n things taken x at a time, also denoted by $C(n,x)$, $_nC_x$, and $\binom{n}{x}$

$$C_x^n = \frac{n!}{x!(n-x)!} = \frac{n(n-1)(n-2) \cdots (n-x+1)}{(1)(2)(3) \cdots (x)}$$

with $0!$ defined as 1.

The probability of getting *at least* r hits in n trials is

$$\sum_{x=r}^{n} P_n(x)$$

The mean and variance of x, the number of successes in n trials, are given by $\mu = np$ and $\sigma^2 = npq$.

Some of the uses of the binomial distribution function follow. It can be used:

1. As a discrete distribution to fit a set of data which are classified in two groups. (When just one sample of size n is available, the binomial distribution indicates how well the unknown parameter p is estimated by the sample proportion x/n. See Sec. 2.3.)

2. As the basic distribution in attributes sampling. (See Chapter 8.)

Tables of the binomial distribution are available (Ref. 1.9 and 2.1).

Example. Suppose that all the shots of a certain series are sure to hit a circular target (Fig. 1.17), and that the probability of a single shot landing in any particular area of the target is proportional to the area. Note that this differs from the normality assumption of Sec. 1.2.1. Then the single-shot probability p of landing in the shaded bull's-eye is the ratio of this area to the total, that is $\pi(1.5)^2/\pi(3)^2 = \frac{1}{4}$. The single shot

TABLE 1.3. BINOMIAL PROBABILITIES FOR
$n = 10$ AND $p = \frac{1}{4}$

No. of bull's-eyes, x	Binomial probability $C_x^{10}(\frac{1}{4})^x(\frac{3}{4})^{10-x}$	Binomial cumulative function
0	0.0563	0.0563
1	0.1877	0.2440
2	0.2816	0.5256
3	0.2503	0.7759
4	0.1460	0.9219
5	0.0584	0.9803
6	0.0162	0.9965
7	0.0031	0.9996
8	0.0004	1.0000
9	0.0000	1.0000
10	0.0000	1.0000

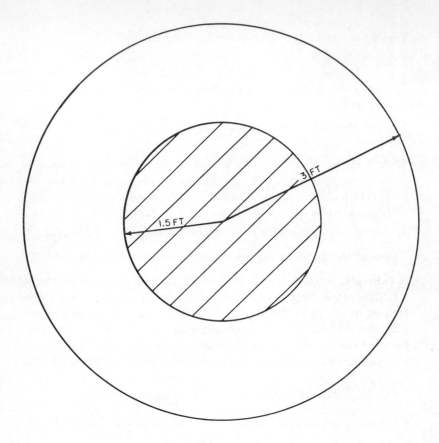

FIG. 1.17.

probability q of missing the bull's-eye is $\frac{3}{4}$. In Table 1.3, the number n of shots is taken as 10. For each entry in the first column, the second column gives the binomial probability of getting exactly that number of bull's-eyes out of the 10 shots. The last column gives the cumulative binomial probability of getting that number of bull's-eyes or fewer. The results are graphed in Fig. 1.18.

1.2.3. Poisson Distribution

As an approximation to the binomial distribution when n is large and p is small, there is the **Poisson distribution**

$$P(x) = \frac{m^x e^{-m}}{x!} \qquad (x = 0, 1, 2, \cdots)$$

where $m = np$ of the approximated binomial distribution and $e = 2.71828$, the base of natural logarithms. Note that $0! = 1$ and $m^0 = 1$.

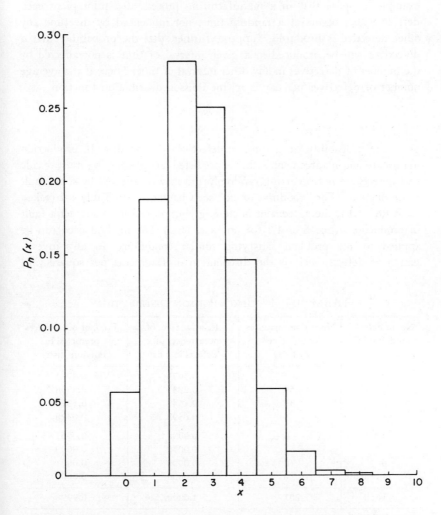

FIG. 1.18. Binomial Distribution for $n = 10$ and $p = \frac{1}{4}$.

The mean and variance of the Poisson distribution are equal: $\mu = m$, and $\sigma^2 = m$. Tables of the Poisson distribution are available (Ref. 1.8).

The Poisson distribution is applicable in some cases in its own right, as well as serving as an approximation to the binomial distribution. For example, suppose that in a manufacturing process at a pilot plant each defective item occurs at a random time, not influenced by the time any other defective is produced. Suppose, further, that the probability that d defectives will be produced in a given interval of time is not affected by the number of defectives in any other interval of time. Then, if the average number of defectives per day is m, the Poisson distribution function

$$P(x) = \frac{m^x e^{-m}}{x!} \qquad (x = 0, 1, 2, \cdots)$$

gives the probability of getting x defectives in one day. If production figures are not in agreement with the predicted frequencies, we must decide that defects are not occurring randomly and may be affected by some fault in production. (For "goodness of fit," see Chapter 4.) In Table 1.4 (taken from Ref. 1.1), the agreement is close, giving no reason to suspect a fault in production. (See Sec. 4.2 for an exact test.) The method used can be applied to any problem satisfying similar conditions. In any investigation of defects, such as the distribution of fragments per square yard

TABLE 1.4. FITTING A POISSON DISTRIBUTION

No. of defects produced in 1 day, x	No. of occurrences of x defects in 1 day	Relative freq. of occurrences of x defects in 1 day	Freq. of occurrences predicted by Poisson distr.
0	102	0.5075	0.4741
1	59	0.2935	0.3538
2	31	0.1542	0.1320
3	8	0.0398	0.0328
4	0	0.0000	0.0061
5	1	0.0050	0.0009
6	0	0.0000	0.0001
7 or more	0	0.0000	0.0000
Total	201	1.0000	0.9998

NOTE: Total number of defects $= 150$; total number of days on which observations were made $= 201$; estimated mean number of defects $= m = 150/201 = 0.74627$.

on a bombarded target, the distribution of flaws per yard of woven sleeving, or the distribution of number of defective igniter squibs per lot, we are interested in whether the occurrence of a flaw is a random result of satisfactory production, or whether it can be ascribed to faulty production.

BIBLIOGRAPHY

1.1. Brownlee, K. A. Industrial Experimentation, 3rd American ed. Brooklyn, N. Y., Chemical Publishing Co., Inc., 1949. Chap. VI, "Poisson Distribution."

1.2. Burington, R. S., and D. C. May, Jr. Handbook of Probability and Statistics With Tables. Sandusky, Ohio, Handbook Publishers, Inc., 1953.

1.3. Dixon, W. J., and F. J. Massey, Jr. An Introduction to Statistical Analysis. New York, McGraw-Hill, 1951. Chap. 1–7 and 14–16.

1.4. Frankford Arsenal. Statistical Manual. Methods of Making Experimental Inferences, 2nd rev. ed., by C. W. Churchman. Philadelphia, Frankford Arsenal, June 1951.

1.5. Hald, A. Statistical Theory With Engineering Applications. New York, Wiley, 1952.

1.6. Kendall, M. G. The Advanced Theory of Statistics, Vol. I, 5th ed. London, Griffin, 1952. Chap. 1, 2, and 8.

1.7. Kendall, M. G., and B. Babington Smith. Tables of Random Sampling Numbers. Cambridge, England, Cambridge University Press, 1939. (Tracts for Computers No. 24.)

1.8. Molina, E. C. Poisson's Exponential Binomial Limit. New York, Van Nostrand, 1942.

1.9. National Bureau of Standards. Tables of the Binomial Probability Distribution. Washington, GPO, 1949. (Applied Mathematics Series 6.)

1.10. U. S. Naval Ordnance Test Station, Inyokern. First Aid for Pet Projects Injured in the Lab or on the Range or What To Do Until the Statistician Comes, by R. M. McClung. China Lake, Calif., NOTS, 29 January 1952. (Technical Memorandum 1113.)

1.11 Wilks, S. S. Elementary Statistical Analysis. Princeton, N. J., Princeton University Press, 1949.

Chapter 2

TESTS AND CONFIDENCE INTERVALS FOR MEANS

2.0. Introduction

This chapter gives tests and confidence intervals for population means based on random samples from those populations. As used here, "test" means a test of a hypothesis, set up prior to the experiment, that the population mean has a specific numerical value. The test yields simply the conclusion that the data are or are not consistent with the hypothesis. The method of testing depends on whether the population standard deviation is known or unknown. In experimentation, the case of unknown standard deviation is the more important and usual; in repetitive work, such as production quality control and standard tests, the standard deviation may sometimes be calculated closely from the large amount of earlier data.

All statistical tests and confidence intervals are based on the assumption that the samples are random, and most of them are based on the assumption that the populations have normal distributions. We are rarely certain that these assumptions are perfectly satisfied. Many large sets of physical and chemical observations have shown at least approximately normal distributions, and experimenters may reasonably feel confident that their variables essentially satisfy the normality assumption on the basis of previously accumulated data of the same type. The results of statistical analysis are in error an unknown amount, depending on the departure of the actual situation from the assumptions. The assumptions of randomness and normality can themselves be objectively tested by methods given in Chapter 4.

What is the importance of the population mean? Suppose we are interested in the weight of a particular type of rocket. If we weigh several of them, we will probably obtain several slightly different weights. Thus, instead of a single weight, we will have a whole population of weights— the weights of all rockets of the particular type considered. In fact, if we make 10 weighings of the same rocket, we will probably note some variation due to errors of measurement. Thus, even for one rocket the exact weight represents only an ideal measurement; actually, there may be a whole population of weights for each rocket—the weights that would be recorded in all possible weighings.

A natural answer to the question, "How much does this rocket weigh?" is the mean of the population of weights that would be recorded in all possible weighings of that particular rocket. A natural answer to the question, "How much do rockets of this type weigh?" is the mean of the population of weights of all such rockets.

Thus, the mean enters into any question of measurement, both because of measurement errors and because of actual variation of the quantity measured. It is, then, highly desirable to have objective methods for testing whether an experimental mean value is consistent with some theoretical value and for stating a confidence interval for the population mean ("true" mean) on the basis of a sample.

If it is practical to take observations successively without specifying beforehand the total sample size, the most economical method for testing a mean is provided by "sequential analysis." (See Sec. 8.4 and its references.)

To find the mean for a quantity that cannot be measured directly, for instance the minimum rocket velocity that will permit penetration of a target, a "staircase" test is often desirable. (See Sec. 4.5 and 4.5.1.)

The τ_1 test in Sec. 2.2.1, the τ_d test in Sec. 2.5.3a, and the sign test in Sec. 2.5.2a are useful for rapid computation, checking, or preliminary tests of means.

2.1. Test for Mean μ When Standard Deviation σ Is Known

2.1.1. Assumptions

(1) The sample is a random sample.

(2) The population is normally distributed with known standard deviation σ—or else the sample size is large (> 30), in which case s may be substituted for σ (see Sec. 3.1).

2.1.2. Normal Test

Given a sample of size n, to test at a given significance level α whether the population mean μ has the hypothetical value a, compute the statistic

$$z = \frac{\bar{x} - a}{\sigma/\sqrt{n}}$$

The distribution of z is normal with unit standard deviation, for $\sigma/\sqrt{n}$ is the standard deviation of means $\bar{x}$.

For an equal-tails test of the null hypothesis that $\mu = a$, reject the hypothesis if z falls in the rejection region $|z| > z_{a/2}$, where $|z|$ is the absolute or numerical value of z. The value $z_{a/2}$ is that exceeded with probability $\alpha/2$ by the normal variable z if the null hypothesis holds.

For a one-sided test of the null hypothesis that $\mu = a$, against the alternative that $\mu > a$, reject the hypothesis if $z > z_a$.

For a one-sided test of the null hypothesis that $\mu = a$, against the alternative that $\mu < a$, reject the hypothesis if $z < -z_a$.

From Appendix Table 2, for a 5% level of significance these rejection regions are

$|z| > 1.960$ for an equal-tails test of $\mu = a$

$z > 1.645$ for a one-sided test of $\mu = a$ against the alternative $\mu > a$

$z < -1.645$ for a one-sided test of $\mu = a$ against the alternative $\mu < a$

If z falls in the rejection region for the test being made, reject the null hypothesis that the population mean μ equals a on the grounds that such a large or small value of z would occur with probability only 5% if the null hypothesis held.

If the population is finite of size N, rather than infinite, replace the above statistic z by

$$z = \frac{\bar{x} - a}{\dfrac{\sigma}{\sqrt{n}} \sqrt{\dfrac{N - n}{N - 1}}}$$

and proceed as before.

Notice that besides the null hypothesis, there are several background assumptions in force: that σ is known, that the statistic z has a normal distribution, and that the sampling was random. Rejection of the null hypothesis may not mean that the hypothesis itself was false; it may mean, instead, that some of the assumptions underlying the construction of the mathematical working model were not satisfied.

On the other hand, failure to reject the null hypothesis is not always strong evidence that it holds. If the sample size is small, only a large departure from the null hypothesis is likely to be adjudged significant. Hence, it is important to plan experiments of sufficient size to hold risks of error to acceptably low levels. Section 2.1.2a describes how to determine suitable sample sizes for testing means when σ is known. Section 1.1.7 may be consulted for a general discussion of statistical tests.

Example. The mean range of a type of rocket is 2,000 yards, and the range standard deviation is 120 yards. Forty rounds fired after a year's storage give a mean range of 1,863 yards. Test at the 5% significance level whether storage changes the mean range.

We compute

$$z = \frac{\bar{x} - a}{\sigma/\sqrt{n}}$$

$$= \frac{1,863 - 2,000}{120/\sqrt{40}} = -7.22$$

For a 5% significance level, the equal-tails rejection region is $|z| > 1.960$. Since -7.22 is less than -1.960, it falls in the rejection region, and we reject the null hypothesis that the mean μ equals 2,000 yards. We conclude that a year's storage does change the mean range of this type of rocket.

An equal-tails rejection region was used under the assumption that *before the experiment was performed* we were interested in a change in range of this rocket in either direction. If we had been interested only in whether the range decreased, we would have used the one-sided rejection region $z < -1.645$.

a. Determination of Sample Size. When an experiment is to be performed and the results are to be analyzed by the method of this section, a decision must be made about the size of the sample to be used. Suppose we wish to guard against the possibility of not rejecting the null hypothesis $\mu = a$ if μ actually has a particular alternative value b. Let there be a specified risk β of making such an error (called a Type II error; see Sec. 1.1.8). We adopt the level of significance α (the risk of rejecting the null hypothesis when the mean actually is a; that is, the probability of a Type I error). In the examples given here, the level α is taken to be 0.05 and the charts included are applicable for that commonly used level only.

From curve A, B, or C of Fig. 2.1, depending upon the type of rejection region used, read off the value of the alternative mean $\mu = a + k\sigma/\sqrt{n}$ for the given risk β. Since the alternative specified to have risk β was $\mu = b$, we have

$$a + \frac{k\sigma}{\sqrt{n}} = b \qquad \text{or} \qquad n = \left(\frac{k\sigma}{b-a}\right)^2$$

Using this sample size assures us that the experiment will enable us to accept the null hypothesis $\mu = a$ or the alternative hypothesis $\mu = b$ with satisfactorily small risks of being wrong in either case.

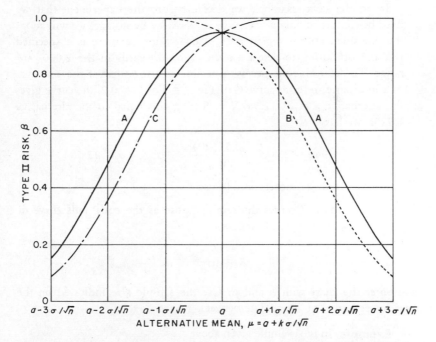

FIG. 2.1. OC Curves for the Normal Test at the 5% Level of Significance. (A) Equal-tails test, (B) one-sided test for an alternative $b > a$, and (C) one-sided test for an alternative $b < a$.

Example. Suppose that in the example given above in this section there is a specified risk 0.2 of failing to detect a change of ± 60 yards in the 2,000-yard range. How large a sample size should be used in a test at the 5% level of significance?

For $\beta = 0.2$ and an equal-tails test, curve A of Fig. 2.1 gives the alternative mean $a \pm 2.8\sigma/\sqrt{n}$. Setting this equal to the alternative $b = 2,000 \pm 60$, we have

$$2,000 \pm 60 = 2,000 \pm \frac{2.8(120)}{\sqrt{n}} \qquad \text{or} \qquad n = 31.4$$

A sample of 32 rockets is required, showing that the sample of 40 actually used had a lower β risk than 0.2 for the alternatives 1,940 and 2,060 yards.

If, in the same example, we decide before the experiment that we need be concerned only if the range is *decreased* by storage, we will adopt the one-sided critical region $z < -1.645$. Suppose there is a specified risk 0.2 of failing to detect a *decrease* of 60 yards in the 2,000-yard range. To determine the necessary sample size for a test at the 5% level of significance, refer to curve C of Fig. 2.1. For $\beta = 0.2$, the curve gives the alternative $\mu = a - 2.5\sigma/\sqrt{n}$. Setting this equal to the alternative 1,940, we have

$$1,940 = 2,000 - \frac{2.5(120)}{\sqrt{n}} \qquad \text{or} \qquad n = 25$$

For this test only 25 rockets need be used.

Appendix Chart V serves the same purpose as the equal-tails curve of Fig. 2.1, but is easier to use. Compute

$$\lambda = \frac{|b - a|}{\sigma}$$

and enter the chart with λ and β. Use the sample size indicated on the next lower curve, or interpolate between curves.

Example. In the example given above

$$\lambda = \frac{|1,940 - 2,000|}{120} = 0.5 \qquad \text{and} \qquad \beta = 0.2$$

In Appendix Chart V, the next lower curve gives 40, but 30 is nearly large enough. Interpolating, we take a sample size of 32 for an equal-tails test.

2.1.3. Confidence Intervals

Sets of $100(1 - \alpha)\%$ confidence limits for the population mean μ are

$$\bar{x} \pm z_{\alpha/2} \frac{\sigma}{\sqrt{n}} \text{ for an infinite population}$$

$$\bar{x} \pm z_{\alpha/2} \frac{\sigma}{\sqrt{n}} \sqrt{\frac{N - n}{N - 1}} \text{ for a finite population of size } N$$

where σ is the known population standard deviation and $z_{\alpha/2}$ is the value in the standard normal distribution such that the probability of a random deviation numerically greater than $z_{\alpha/2}$ is α. (See Sec. 1.1.10 for a general discussion of confidence intervals.)

Example. In an assessment process several measurements are made on each photographic plate to determine the distance between two points. For one plate the average of four measurements is 1.215 inches. Assume that the population of measurements has a normal distribution with standard deviation 0.01 inch. What is a 95% confidence interval for the distance between the points?

From Appendix Table 2, we find $z_{.025} = 1.960$. Using the first formula of this section, we obtain

$$1.215 \pm \frac{1.960\,(0.01)}{\sqrt{4}}$$

This gives the 95% confidence interval of 1.205 to 1.225 inches.

a. Determination of Sample Size. We can determine the necessary sample size so that the confidence interval will be of length $2h$ (giving confidence limits $\bar{x} \pm h$). From

$$2z_{\alpha/2} \frac{\sigma}{\sqrt{n}} = 2h$$

we have

$$n = \left(\frac{z_{\alpha/2}\sigma}{h}\right)^2$$

Example. If in the example above we specify the length of a 95% confidence interval to be 0.015 inch, how many measurements should be made on each plate?

We compute

$$n = \left[\frac{(1.960)(0.01)}{0.0075} \right]^2 = 6.8$$

A sample size of seven is required.

2.2. Test for μ When σ Is Unknown

Two useful tests are given below for this case—the τ_1 test and the t test. The first is quickly and easily applied, but is restricted to small samples ($2 \leqslant n \leqslant 10$) and should not be used if the population is known, or suspected, to be non-normal. It is often used for making a preliminary check on a portion of a larger sample. The t test involves considerably more computation, but is the most powerful test available for this case. It is valid for any sample size, and it is a good approximation, especially for large samples, even though the population is non-normal to a noticeable degree.

2.2.1. The τ_1 Test

To apply a one-sided τ_1 test of the null hypothesis $\mu = a$, against the alternative $\mu > a$, on the basis of a random sample from a normal population, compute

$$\tau_1 = \frac{\bar{x} - a}{w}$$

where w is the range of the sample. Compare τ_1 with the value in Appendix Table 13(a) for the given sample size and the desired level of significance. Reject the null hypothesis that $\mu = a$ if τ_1 exceeds the tabled value (Ref. 2.2). The equal-tails test is identical except that the sign of τ_1 is ignored and the critical value used is that for $\alpha/2$.

2.2.2. The *t* Test

For a *t* test of the null hypothesis $\mu = a$ on the basis of a random sample, compute

$$t = \frac{\bar{x} - a}{s/\sqrt{n}}$$

where

$$s^2 = \frac{1}{n-1} \sum (x_i - \bar{x})^2 = \frac{1}{n(n-1)} \left[n \sum x_i^2 - (\sum x_i)^2 \right]$$

as in the definition of Sec. 1.1.4a. For an equal-tails test, compare the calculated value with the value $t_{a/2, n-1}$ under $P(t) = \alpha/2$ in Appendix Table 3. The subscript $n-1$ represents the number of degrees of freedom f. If $|t| > t_{a/2, n-1}$, reject the null hypothesis that $\mu = a$.

For a one-sided test of the null hypothesis $\mu = a$, against the alternative $\mu > a$, use the rejection region $t > t_{a, n-1}$.

Example. The following readings represent the measurements of outside diameter x_i (in inches) for nine grains of the same type.

2.021	2.002	2.001
2.005	1.990	1.990
2.009	1.983	1.987

Assuming a normal distribution, we wish to test at the 5% significance level whether the average diameter of this type of grain is 2.000 inches.

To simplify computation we code the data, replacing each x_i by $y_i = 1,000 \, (x_i - 2.000)$. Then the computations are performed on the y_i's and

$$\bar{x} = \frac{\bar{y} + 2,000}{1,000}, \qquad s_x = \frac{s_y}{1,000}$$

We now have the following y_i's:

21	2	1
5	-10	-10
9	-17	-13

We compute

$$\sum y_i = -12$$

$$\bar{y} = -1.333$$

$$\bar{x} = 2 + (-0.001333)$$

$$\sum y_i^2 = 1,210$$

$$s_y^2 = \frac{1}{9 \cdot 8}\left[(9)(1,210) - (-12)^2\right] = 149.25$$

$$s_y = 12.22$$

$$s_x = 0.01222$$

$$t = \frac{2 + (-0.001333) - 2}{0.01222/3} = -0.327$$

From Appendix Table 3 we find $t_{.025,8} = 2.306$. Since this is greater than the absolute value of our computed t, we cannot reject the null hypothesis that the average diameter of this type of grain is 2.000 inches.

The statistic t can also be computed directly from the coded values:

$$t = \frac{\bar{y} - 0}{s_y/\sqrt{n}} = \frac{-1.333}{12.22/3} = -0.327$$

In this example, since $2 \leqslant n \leqslant 10$, the less powerful τ_1 test could have been used as an alternative procedure.

a. Determination of Sample Size. An experiment is to be performed and the results are to be analyzed by the equal-tails t test described above. As in Sec. 2.1.2a, we can determine the necessary sample size for a t test to be conducted at the significance level α if there is specified a risk β of accepting the null hypothesis $\mu = a$ when actually μ has the alternative value b. It is necessary to have an estimate of the standard deviation (obtained from past experience with similar data).

For a significance level $\alpha = 0.05$, Appendix Chart VI can be used to obtain the sample size. Compute

$$\lambda = \frac{|b - a|}{\sigma}$$

using for σ an estimate of the standard deviation. Opposite λ and β in Appendix Chart VI, select the sample size n indicated on the next lower curve, or interpolate between curves.

The absolute (numerical) value of $|b - a|$ is used since Appendix Chart VI is for an equal-tails test.

Example. In the case cited above, find the sample size necessary to test the hypothesis $\mu = 2.000$ inches at the 5% level of significance. Assume that σ is thought from past experience to have a value between 0.005 and 0.010 inch, and that a β risk of 0.2 is specified for the error of accepting the null hypothesis $\mu = 2.000$ when actually μ has an alternative value of 2.000 ± 0.010.

Using the two extreme values for σ, we obtain

$$\lambda_1 = \frac{|2.000 - 2.010|}{0.005} = 2$$

$$\lambda_2 = \frac{|2.000 - 2.010|}{0.010} = 1$$

For $\lambda = 2$ and $\beta = 0.2$, Appendix Chart VI gives $n = 5$. For $\lambda = 1$ and $\beta = 0.2$, the chart gives $n = 10$. Thus, to be sure of keeping the probability of a Type II error as small as specified, a sample size of 10 should be used.

Notice that no matter how small a sample we have (provided that $n > 1$, so that s can be calculated), we can test the hypothesis $\mu = a$ at any significance level, say $\alpha = 0.001$ or even less. Thus we make sure of avoiding the error of rejecting the hypothesis $\mu = a$ when it is true; but we are likely to miss recognizing an alternative situation $\mu = b$ when *it* is true. The only way to make the risks of both errors sufficiently small is to choose a sample size that is sufficiently large.

2.2.3. Confidence Intervals

Confidence limits for the mean μ with coefficient $100(1 - \alpha)\%$ are

$$\bar{x} \pm t_{\alpha/2,\,n-1} \frac{s}{\sqrt{n}}$$

where $t_{a/2,n-1}$ is again the t deviate for $f = n - 1$ degrees of freedom, the probability of exceeding which is $P(t) = \alpha/2$.

Example. A recently developed fuze is tested for the first time at high temperatures. The coded mean of arming distances for 10 fuzes is 42 units of distance. The value of s is 2.1. Under the supposition that the arming distances have a normal distribution, find a 95% confidence interval for the mean arming distance of the new fuze.

From Appendix Table 3 we find $t_{.025,9} = 2.262$. From $\bar{x} \pm t_{a/2,n-1}s/\sqrt{n}$ we have $42 \pm 2.262\,(2.1)/\sqrt{10}$. The desired confidence interval is 40.5 to 43.5 units of distance.

a. Determination of Sample Size. If an estimate of σ is available, the equation

$$n = \left(\frac{\sigma t_{a/2,n-1}}{h}\right)^2$$

can be used to give the sample size necessary to obtain a $100(1 - \alpha)\%$ confidence interval with *expected* length $2h$. The *actual* length obtained from the data will be calculated from the first equation in this section; therefore it will depend on the s calculated from the sample, varying randomly above or below $2h$ to an extent depending on the discrepancy between σ and s.

Example. How many readings of temperature should be taken if it is hoped to specify the temperature within $\pm 0.5°$ with 99% confidence? From previous trials made with the same thermometer and the same observer, the standard deviation of the readings is expected to be in the neighborhood of $0.6°$.

In this case $\sigma = 0.6°$ and $h = 0.5°$. Using Appendix Table 3, we find by trial and error that for $n = 13$, $(\sigma t_{.025,n-1}/h)^2 = 13.44$; and for $n = 14$, $(\sigma t_{.025,n-1}/h)^2 = 13.06$. This shows that for $n = 13$ the expected length of the confidence interval will be slightly more than the prescribed value $0.5°$, and for $n = 14$ it will be slightly less. In such a situation, the usual practice is to choose the larger sample size, $n = 14$.

2.3. Test for Proportion p

Often the sample items are merely classified as having or not having a certain characteristic; e.g., they function or fail to function, they succeed or fail, they are perfect or contain defects. Here the basic distribution is the binomial (Sec. 1.2.2). We are interested in testing a hypothetical value p_0 for the population proportion having the characteristic. The unknown parameter p is estimated from a sample as the ratio r/n of the number r of items having the characteristic to the sample size n.

2.3.1. Assumptions

(1) A proportion $p \leqslant 0.5$ of the items of a population have a particular characteristic. The artificial restriction $p \leqslant 0.5$, which is a convenience in constructing tables, is satisfied by considering p as the proportion *not* having the characteristic, if the proportion *having* the characteristic is greater than 0.5.

(2) The sample is a random sample, each item of which is tested for the characteristic.

(3) Ideally, the population is infinite. If the population is finite (as in a pilot lot), the methods described in Sec. 2.3.2 give good approximations for lot sizes down to 10 times the sample size. (For finite populations, see also the end of Sec. 2.3.3b.)

2.3.2. Test

a. Test for a Sample of Size $n \leqslant 150$. To test at the 5% level the hypothesis that p has the value p_0, against alternative values greater than p_0, find the smallest whole number r for which

$$\sum_{s=r}^{n} C_s^n \, p_0^s (1 - p_0)^{n-s} \leqslant 0.05$$

using one of the tables described in the next paragraph. Reject the null hypothesis if r or more defectives are observed in the sample. Note that this is a one-sided test.

The technique shown here is correct for any sample size, though its practical use depends on the availability of tables. Table 2 of Ref. 2.5 lists the values of the entire summation shown above for $n < 50$, for all values of r, and for p_0 from 0.01 to 0.50 in steps of 0.01. Ref. 2.7 extends the table of Ref. 2.5 from $n = 50$ to $n = 100$ in steps of 5, and Ref. 2.1 expands the table to include all $n \leqslant 150$.

Example. A sample of 40 rockets is chosen at random from a large lot. Design a test for determining at the 5% level of significance whether the proportion of rockets defective (in that they fail to arm) is 0.01, assuming that we are interested in detecting an alternative proportion greater than 0.01.

Using Table 2 of Ref. 2.5, we find

$$\sum_{s=3}^{40} C_s^{40} (0.01)^s (0.99)^{40-s} = 0.007$$

$$\sum_{s=2}^{40} C_s^{40} (0.01)^s (0.99)^{40-s} = 0.061$$

Thus, if the sample contains three or more defectives, the lot should be rejected as having a proportion defective greater than 0.01.

b. Test for a Sample of Size $n > 5/p_0$. To test at the significance level α the hypothesis that $p = p_0$ against the alternative that $p > p_0$, compute

$$z = \frac{r - np_0 - \frac{1}{2}}{\sqrt{np_0(1 - p_0)}}$$

where r is the number of defectives observed in a sample of n. Compare z with z_α (from Appendix Table 2 with $F(z) = 1 - \alpha$) and reject the hypothesis if $z > z_\alpha$.

To test the null hypothesis $p = p_0$ against the alternative that $p < p_0$, compute

$$z = \frac{r - np_0 + \frac{1}{2}}{\sqrt{np_0(1 - p_0)}}$$

and reject the null hypothesis if $z < -z_\alpha$.

For an equal-tails test of the null hypothesis against alternatives greater or less than p_0, compute

$$z = \frac{|r - np_0| - \frac{1}{2}}{\sqrt{np_0(1 - p_0)}}$$

and reject the null hypothesis if $z > z_{\alpha/2}$.

These tests are based on a normal approximation that is sufficiently good for $n > 5/p_0$. For a sequential test for a proportion p, see Sec. 8.4.1.

c. Determination of Sample Size. The method for determining the required sample size will not be taken up explicitly for the testing of hypotheses. It is convenient to use the sample size n necessary to give a confidence interval of given length (Sec. 2.3.3a) for the significance test as well.

2.3.3. Confidence Intervals

a. Graphical and Tabular Solutions. Appendix Charts II, III, and IV give two-sided 90, 95, and 99% confidence limits, respectively, for p as a function of the proportion defective in the sample for a number of sample sizes. For $n \leqslant 30$, Appendix Table 21(a) should be used. The confidence coefficient is in each case at least as high as indicated; it cannot be made exact because of the discrete nature of the distribution. The charts can also be used to determine roughly the necessary sample size for a confidence interval of specified length. (See the second example below.)

For *one-sided* confidence intervals, use Appendix Table 21(b). Appendix Charts II, III, and IV can be used for one-sided intervals with confidence coefficients of 95, 97.5, and 99.5%, respectively, by referring to only one of a symmetrical pair of curves.

Example. The proportion defective in a random sample of 250 items checked on a go–no-go gage is 12%. What are 95% confidence limits for the proportion defective for total production?

The desired confidence limits are marked off by the two curves for $n = 250$ in Appendix Chart III at $r/n = 0.12$. We obtain a confidence interval of 8 to 17%.

Example. In estimating the proportion defective, the maximum allowable length for a 95% confidence interval is 0.10. What sample size should be used?

From Appendix Chart III, we find that the 95% confidence interval has maximum length 0.14 for $n = 250$, and 0.06 for $n = 1,000$. Linear interpolation with respect to $1/n$ should be used.

n	$1,000/n$	*Maximum length*
250	4 ⎫	0.14 ⎫
400	2.5 ⎬ 3	0.10 ⎬ 0.04 ⎫ 0.08
1,000	1 ⎭	0.06 ⎭

$$4 - \frac{0.04}{0.08}\,(3) = 2.5$$

We find, then, $n = 1,000/2.5 = 400$.

It is perhaps surprising to find that such a large sample size is required to locate a proportion with an interval of length not more than 0.10 with 95% confidence; however, the significance of a proportion from a small sample is often overrated. For example, if we test a sample of 10 new missiles and all 10 fly properly, we may feel confident of the missile, but the 95% confidence limits for the proportion defective are 0 and 0.267.

b. Numerical Solution. For a large sample, $100\,(1 - \alpha)\%$ confidence limits for p are given by

$$\frac{r + z_{a/2}^2/2 \pm \sqrt{(r + z_{a/2}^2/2)^2 - (n + z_{a/2}^2)\,r^2/n}}{n + z_{a/2}^2}$$

where r is the number of defectives observed and $z_{a/2}$ is the normal deviate exceeded with probability $\alpha/2$. In case the population is finite of size N, replace $z_{a/2}^2$ by $z_{a/2}^2\,(N - n)/(N - 1)$.

2.4. Test for $\mu_1 - \mu_2$ When σ_1 and σ_2 Are Known

2.4.1. Test Conditions and Assumptions

A random sample is drawn from each of two populations to determine whether the difference between the two population means μ_1 and μ_2 is equal to d. The two samples may be independent, or the observations may be made in pairs, one from each population, extraneous variables being held constant for each pair. Let y denote the difference $x_1 - x_2$ of the paired observations. If the standard deviation σ_y resulting from such pairing is known and is less than $\sqrt{\sigma_{x_1}^2 + \sigma_{x_2}^2}$, then it is desirable to pair during sampling. This situation will arise in general if the paired observations are positively correlated. (See Sec. 2.5 for discussion of choice of method if population standard deviations are unknown.)

Each population is assumed to be normally distributed with known standard deviation σ_i—or else the sample drawn from it is large (> 30), in which case s_i may be substituted for σ_i (see Sec. 3.1).

2.4.2. Normal Test

a. Test for Paired Observations. Subtract the second reading of each pair from the first and consider the resulting differences as sample values of a new variable y. Test the null hypothesis that $\mu_y = d$ by the method of Sec. 2.1 if σ_y is known, or by the method of Sec. 2.2 if σ_y is unknown. (It is known from theory that $\mu_y = \mu_1 - \mu_2$.) Confidence intervals for μ_y and determinations of sample size can also be obtained by the methods of Sec. 2.1 and 2.2.

b. Test for Two Independent Samples. To test at the significance level α the hypothesis that $\mu_1 - \mu_2 = d$, compute

$$z = \frac{\bar{x}_1 - \bar{x}_2 - d}{\left(\dfrac{\sigma_1^2}{n_1} + \dfrac{\sigma_2^2}{n_2}\right)^{\frac{1}{2}}}$$

As in Sec. 2.1.2, reject the null hypothesis that $\mu_1 - \mu_2 = d$ if z falls in the rejection region $|z| > z_{\alpha/2}$ for an equal-tails test or $z > z_\alpha$ for a one-sided test against alternatives greater than d.

If the two populations are finite of sizes N_1 and N_2, replace σ_1^2 by $\sigma_1^2(N_1 - n_1)/(N_1 - 1)$ and σ_2^2 by $\sigma_2^2(N_2 - n_2)/(N_2 - 1)$ in the formula for z.

Example. A standard method of chemical analysis for determining nitrocellulose in propellants is known, from the large number of previous analyses, to have a standard deviation $\sigma = 0.6$. Each of two new propellants, assumed homogeneous, was subjected to five analyses. One result was lost because of an accident; the other nine results were:

$$x_{1j}: \quad 63.12, \ 63.57, \ 62.81, \ 64.32, \ 63.76\%$$

$$x_{2j}: \quad 62.54, \ 63.21, \ 62.38, \ 62.06\%$$

We wish to test at the 5% significance level whether there is any real difference in nitrocellulose content.

The hypothesis to be tested is that $\mu_1 = \mu_2$, or $\mu_1 - \mu_2 = 0$. We compute

$$\bar{x}_1 = 62 + \frac{7.58}{5} = 63.516$$

$$\bar{x}_2 = 62 + \frac{2.19}{4} = 62.548$$

Then

$$z = \frac{\bar{x}_1 - \bar{x}_2 - 0}{\left(\dfrac{\sigma^2}{n_1} + \dfrac{\sigma^2}{n_2}\right)^{\frac{1}{2}}} = \frac{0.968}{0.6\left(\dfrac{1}{5} + \dfrac{1}{4}\right)^{\frac{1}{2}}} = 2.405$$

From Appendix Table 2, for $F(z) = 1 - \alpha/2$ we find $z_{.025} = 1.960$. Since 2.405 is greater than 1.960, we reject the null hypothesis that $\mu_1 = \mu_2$ at the 5% level of significance.

c. Determination of Size of Independent Samples. To make most effective use of a fixed number N of items allocated for sampling, choose

$$n_1 = \frac{\sigma_1}{\sigma_1 + \sigma_2} N \quad \text{and} \quad n_2 = N - n_1$$

If $\sigma_1 = \sigma_2 = \sigma$, choose $n_1 = n_2 = n$. To determine n for an equal-tails test at the 5% level of significance with an assigned risk β of accepting the hypothesis that $\mu_1 - \mu_2 = d$ when it really has an alternative value b, compute

$$\lambda = \frac{|b - d|}{\sigma\sqrt{2}}$$

and refer to Appendix Chart V.

Example. A delayed-action fuze in present use has a standard deviation in delay time of 0.004 seconds. An experiment is to be performed to determine whether a new experimental fuze has the same mean delay time. The test is designed to detect at the 5% significance level either an increase or a decrease. The standard deviation for the two fuzes is thought to be the same. How large a sample of each should be used, if there is an assigned risk of 0.05 of deciding that $\mu_1 - \mu_2 = 0$ when $|\mu_1 - \mu_2|$ has the alternative value of 0.005?

Here $d = 0$, $b = 0.005$, and $\beta = 0.05$. To detect a change in delay time in either direction, an equal-tails test will be used. Since $\sigma_1 = \sigma_2 = 0.004$, we choose $n_1 = n_2 = n$. We have

$$\lambda = \frac{|0.005 - 0|}{(0.004) \sqrt{2}} = 0.88$$

From Appendix Chart V for $\lambda = 0.88$, $\alpha = 5\%$, and $\beta = 5\%$, by interpolating between curves we obtain the sample size $n = 18$. Thus, 18 fuzes of each type should be tested.

2.4.3. Confidence Intervals

A set of $100(1 - \alpha)\%$ confidence limits for $\mu_1 - \mu_2$ for independent samples is

$$(\bar{x}_1 - \bar{x}_2) \pm z_{\alpha/2} \left(\frac{\sigma_1^2}{n_1} + \frac{\sigma_2^2}{n_2} \right)^{\frac{1}{2}}$$

a. Determination of Size of Independent Samples. If the length $2z_{\alpha/2} [(\sigma_1^2/n_1) + (\sigma_2^2/n_2)]^{\frac{1}{2}}$ of the $100(1 - \alpha)\%$ confidence interval for $\mu_1 - \mu_2$ is specified as $2h$, the total sample size should be

$$N = n_1 + n_2 = \left[\frac{z_{\alpha/2}(\sigma_1 + \sigma_2)}{h} \right]^2$$

giving

$$n_1 = \left(\frac{z_{\alpha/2}}{h} \right)^2 \sigma_1 (\sigma_1 + \sigma_2) \quad \text{and} \quad n_2 = \left(\frac{z_{\alpha/2}}{h} \right)^2 \sigma_2 (\sigma_1 + \sigma_2)$$

2.5. Test for $\mu_1 - \mu_2$ When σ_1 and σ_2 Are Unknown

2.5.1. Assumptions and Design of Experiment

The assumptions vary with the different situations and tests given below. Random samples are necessary, as usual. Normal distributions are assumed except in the sign test. Using the common normality assumption, we must decide whether to sample the two populations independently, or to pair observations so that conditions which usually fluctuate are the same for each pair, though changing from pair to pair. Thus, in a durability test of paints, panels of two or more paints would be exposed side by side and would thus give paired rather than independent observations. If the conditions that are rendered the same by pairing are known to influence substantially the characteristic x under observation, then it is probably desirable to pair observations. The advantage of independent samples lies in the doubling of the number of degrees of freedom with no increase in the number of observations, but this advantage may or may not be overbalanced by the reduction in variance afforded by pairing; each problem must be examined carefully. Pairing has the additional advantage of eliminating any assumption about the equality of σ_1 and σ_2.

2.5.2. Test for Paired Observations

a. Sign Test. (See Ref. 2.2.) The sign test provides a quick and simple way of determining whether a set of paired readings shows a significant mean difference between pair members. Unlike the t test described below, this test does not depend on an assumption of normality; however, when normality assumptions are valid, it is not as efficient as the t test. For the sign test, it is unnecessary that the different pairs be observed under the same conditions, so long as the two members of each pair are alike. As used here, it is an equal-tails test, but it can be adapted for use as a one-sided test.

Example. The first two lines of the tabulation below represent the deflection in mils of rockets fired in pairs simultaneously from two launchers, A and B. Can we say that there is a difference between the performances of the two launchers at the 5% significance level?

To obtain the signs shown, we record a "plus" whenever the A row shows a higher value than the B row, a "minus" when it is lower, and a zero when it is the same.

A	4.4	−1.4	3.2	0.2	−5.0	0.3	1.2	2.2	1.3	−0.7
B	3.2	7.7	6.4	2.7	3.1	0.6	2.6	2.2	2.2	0.9
Sign	+	−	−	−	−	−	−	0	−	−

We note that there are fewer +'s than −'s, that the total number N of +'s and −'s is 9, and that the number of +'s is 1. The zero is omitted from all counts. In Appendix Table 9 opposite 9 under $\alpha = .05$, we find the number 1. Since the number of + signs is *less than or equal to the tabled value 1*, we decide that there is a significant difference between the performances of the two launchers at the 5% level.

b. The t Test on Differences. The t test, which depends on an assumption of normality and requires that all pairs be observed under the same conditions, gives more information than the sign test. Subtract the pair values and consider the resulting differences as the data to be analyzed by the method of Sec. 2.2.2.

2.5.3. Test for Two Independent Samples When $\sigma_1 = \sigma_2$

a. The τ_d Test. For small samples $(n_1 = n_2 = n \leqslant 10)$ or for rapid pilot computations using part of a large sample, compute

$$\tau_d = \frac{\bar{x}_1 - \bar{x}_2}{w_1 + w_2}$$

where the w's are the respective ranges. Compare τ_d with Appendix Table 13(b) and reject at the significance level α the hypothesis that the means are equal if τ_d exceeds the value tabled under α. (See Ref. 2.2.) This is a one-sided test, but the equal-tails test is identical except that the sign of τ_d is ignored and the critical value used is that for $\alpha/2$.

b. The t Test. In general, to test the hypothesis that $\mu_1 - \mu_2 = d$, compute

$$t = \frac{\bar{x}_1 - \bar{x}_2 - d}{s_0 \left(\dfrac{1}{n_1} + \dfrac{1}{n_2} \right)^{1/2}}$$

where

$$s_0^2 = \frac{\sum\limits_{j=1}^{n_1} (x_{1j} - \bar{x}_1)^2 + \sum\limits_{j=1}^{n_2} (x_{2j} - \bar{x}_2)^2}{n_1 + n_2 - 2}$$

is a "pooled" estimate (see Sec. 3.1). Compare t with the value $t_{\alpha/2, n_1+n_2-2}$ in Appendix Table 3, and reject the hypothesis if t exceeds the tabled value. As in Sec. 2.2.2 for an equal-tails test, the value of $t_{\alpha/2, n_1+n_2-2}$ is the t value for $f = n_1 + n_2 - 2$ degrees of freedom such that the probability is $\alpha/2$ of a random reading greater than $t_{\alpha/2, n_1+n_2-2}$ and $\alpha/2$ of a reading less than $-t_{\alpha/2, n_1+n_2-2}$. For a one-sided test, compare t with t_{α, n_1+n_2-2}.

To determine sample size for an equal-tails test at the 5% significance level with an assigned risk β of accepting the hypothesis $\mu_1 - \mu_2 = d$ when $\mu_1 - \mu_2$ actually has the alternative value b, compute

$$\lambda = \frac{|d - b|}{\sigma \sqrt{2}}$$

For σ, use an estimate of the common standard deviation of the two populations. Refer to Appendix Chart VI for the common sample size $n_1 = n_2 = N/2$.

Confidence limits for $\mu_1 - \mu_2$ with confidence coefficient $1 - \alpha$ are

$$(\bar{x}_1 - \bar{x}_2) \pm t_{\alpha/2, n_1 + n_2 - 2} s_0 \left(\frac{1}{n_1} + \frac{1}{n_2} \right)^{\frac{1}{2}}$$

Example. The variability of two cutting machines is assumed to be the same. One is set to cut pieces 3.00 mm shorter than the other to allow for insertion of a compressible spacer. There is assigned a risk of 0.05 of accepting the hypothesis that the mean difference is 3.00 mm when it has an alternative value 3.00 ± 1.00 mm. If the standard deviation in both cases is of the order of magnitude of 0.5 mm, what sample size should be used for a test at the 5% level of significance?

Taking $n_1 = n_2 = n$, we have

$$\lambda = \frac{|d - b|}{\sigma \sqrt{2}} = \frac{|1.00|}{0.5 \sqrt{2}} = 1.41$$

TABLE 2.1. Length in mm of Pieces Produced
by Two Cutting Machines

Machine 1, x_{1j}	Machine 2, x_{2j}
98.06	94.73
98.07	95.02
98.93	94.36
98.03	96.16
98.50	94.50
98.60	94.82
99.67	93.75
98.51	93.90
98.06	95.31
99.01	95.21

From Appendix Chart VI for an equal-tails test with $\lambda = 1.41$ and $\beta = 0.05$, we read $n = 10$ for each machine.

Suppose that when the experiment is run, the readings in mm are as shown in Table 2.1. After coding the readings by replacing x_{1j} by $y_{1j} = 100 (x_{1j} - 98)$ and x_{2j} by $y_{2j} = 100 (x_{2j} - 93)$, the data will appear as shown in Table 2.2.

TABLE 2.2. CODED LENGTHS OF PIECES
FROM TWO CUTTING MACHINES

y_{1j}	y_{2j}
6	173
7	202
93	136
3	316
50	150
60	182
167	75
51	90
6	231
101	221

We compute

$$\sum y_{1j} = 544$$

$$\bar{x}_1 = 98 + 0.544 = 98.544$$

$$\sum y_{1j}^2 = 55,570$$

$$\sum (y_{1j} - \bar{y}_1)^2 = \sum y_{1j}^2 - \frac{1}{n} \left(\sum y_{1j}\right)^2 = 25,976.4$$

$$\sum y_{2j} = 1,776$$

$$\bar{x}_2 = 93 + 1.776 = 94.776$$

$$\sum y_{2j}^2 = 360,636$$

$$\sum (y_{2j} - \bar{y}_2)^2 = \sum y_{2j}^2 - \frac{1}{n} \left(\sum y_{2j}\right)^2 = 45,218.4$$

$$s_{0y}^2 = \frac{25,976.4 + 45,218.4}{18} = 3,955.267$$

$$s_{0y} = 62.891$$

$$s_{0x} = 0.6289$$

$$t = \frac{98.544 - 94.776 - 3}{0.6289\left(\dfrac{1}{10} + \dfrac{1}{10}\right)^{\frac{1}{2}}} = 2.730$$

$$t_{.025,18} = 2.101$$

We reject the null hypothesis that the mean difference between the lengths cut by Machine 1 and those cut by Machine 2 is 3 mm.

2.5.4. Test of $\mu_1 - \mu_2 = 0$ for Two Independent Samples When $\sigma_1 \neq \sigma_2$

For a general treatment of this subject, see Ref. 2.6, p. 91.

Label the samples so that $n_1 \leqslant n_2$. Use the readings in the order of observation or else randomize them by lottery choice. It is best in designing such an experiment to make the two sample sizes equal, since the analysis makes little use of extra sample items from either population.

Define

$$u_i = x_{1i} - x_{2i}\sqrt{\frac{n_1}{n_2}} \qquad (i = 1, 2, \cdots, n_1)$$

$$\bar{u} = \frac{1}{n_1}\sum_{i=1}^{n_1} u_i$$

$$Q = n_1\sum_{i=1}^{n_1}(u_i - \bar{u})^2 = n_1\sum_{i=1}^{n_1}u_i^2 - (\sum_{i=1}^{n_1}u_i)^2$$

Then compute

$$t = \frac{\bar{x}_1 - \bar{x}_2}{\sqrt{\dfrac{Q}{n_1^2(n_1-1)}}}$$

where

$$\bar{x}_1 = \frac{1}{n_1}\sum_{i=1}^{n_1} x_{1i}, \qquad \bar{x}_2 = \frac{1}{n_2}\sum_{i=1}^{n_2} x_{2i}$$

Reject the null hypothesis that $\mu_1 = \mu_2$ at the significance level α if t falls in the rejection region $|t| > t_{\alpha/2,n_1-1}$ for an equal-tails test or the region

$t > t_{a,n_1-1}$ for a one-sided test. Appendix Table 3 gives percentage points for the t distribution.

Example. Range data are given in Table 2.3 for two types of rockets. The true (population) range dispersions are not expected to be equal.

TABLE 2.3. RANGE DATA FOR TWO TYPES OF ROCKETS

Type II, x_{2i}, yd	Type I, x_{1i}, yd	$x_{2i}\sqrt{n_1/n_2}$	$u_i = x_{1i} - x_{2i}\sqrt{n_1/n_2}$
1,472	1,747	1,343.7	403.3
1,315	1,618	1,200.4	417.6
1,984	1,521	1,811.1	−290.1
1,900	1,137	1,734.5	−597.5
1,646	1,374	1,502.6	−128.6
1,709	1,325	1,560.1	−235.1
1,780	1,821	1,624.9	196.1
1,571	2,351	1,434.1	916.9
1,681	1,883	1,534.5	348.5
1,453	1,613	1,326.4	286.6
1,799	1,843	1,642.3	200.7
1,372	1,796	1,252.5	543.5
1,420	1,507	1,296.3	210.7
1,612	1,387	1,471.5	−84.5
1,546	725	1,411.3	−686.3
1,255	1,041	1,145.7	−104.7
1,818	1,652	1,659.6	−7.6
1,357	1,595	1,238.8	356.2
1,412	1,679	1,289.0	390.0
1,775	1,557	1,620.3	−63.3
1,850	1,206	1,688.8	−482.8
1,417	192	1,293.5	−1,101.5
2,057	1,025	1,877.8	−852.8
1,468	813	1,340.1	−527.1
1,694	971	1,546.4	−575.4
1,776	....	....	....
1,456	....	....	....
1,489	....	....	....
1,618	....	....	....
1,544	....	....	....

Test the mean ranges of the two types of rockets for equality at the 5% significance level.

We compute

$$\bar{x}_2 = 1,608.2 \qquad\qquad \bar{x}_1 = 1,415.2$$

$$n_2 = 30 \qquad\qquad n_1 = 25$$

$$\sum_{i=1}^{n_1} u_i = -1,467.2 \qquad\qquad \sum_{i=1}^{n_1} u_i^2 = 5,867,573.36$$

$$Q = 25(5,867,573.36) - (-1,467.2)^2 = 144,536,658.16$$

$$\frac{Q}{n_1^2(n_1 - 1)} = 9,635.78$$

$$\left[\frac{Q}{n_1^2(n_1 - 1)}\right]^{\frac{1}{2}} = 98.16$$

$$t = \frac{\bar{x}_1 - \bar{x}_2}{\left[\dfrac{Q}{n_1^2(n_1 - 1)}\right]^{\frac{1}{2}}} = -\frac{193.0}{98.16} = -1.966$$

From Appendix Table 3, $t_{a/2,n_1-1} = t_{.025,24} = 2.064$. Since 1.966 does not exceed 2.064, the hypothesis that the two means are equal cannot be rejected at the 5% significance level.

2.6. Test for a Difference Between Proportions, $p_1 - p_2$

To determine whether two samples come from binomial populations having the same proportion defective, see Sec. 4.7.1 on contingency tables.

2.7. Test for Differences Among Several Means

Use the method of "analysis of variance" as described in Chapter 5.

2.8. Detection of a Trend in a Set of Means, Standard Deviations Unknown but Assumed Equal

In addition to the sections below, see Chapter 6.

2.8.1. Assumptions

A random sample of size n is drawn from each of k normal populations with the same standard deviation. The common sample size n may be 1.

2.8.2. Method

A trend in means is some relationship among the means which can be represented graphically by either a straight line or a curve. The ordinary formula for variance among means in a k sample problem is

$$s^2 = \frac{1}{k-1} \sum_{i=1}^{k} (\bar{x}_i - \bar{\bar{x}})^2$$

where $\bar{\bar{x}}$ is the mean of the k sample means and $\bar{x}_i$ is the mean of the ith sample. Now s^2 will include the effect of a trend in means, if a trend is present. An estimate of variance which reduces the trend effect is $\frac{1}{2}\delta^2$ where

$$\delta^2 = \frac{1}{k-1} \sum_{i=1}^{k-1} (\bar{x}_{i+1} - \bar{x}_i)^2$$

$$= \textit{mean square successive difference}$$

To test whether a trend in means is present, we compute δ^2/s^2 and compare the result with the appropriate critical value of δ^2/s^2 in Appendix Table 14.

TABLE 2.4. GLUCOSE CONCENTRATION OF FIVE
MIXTURES TESTED OVER A 35-WEEK PERIOD

No. of weeks	Mean % glucose, $\bar{x}_i$	$\bar{x}_{i+1} - \bar{x}_i$
4	43.14	
		−0.98
8	42.16	
		2.84
12	45.00	
		−1.00
16	44.00	
		3.86
20	47.86	
		0.29
24	48.15	
		1.95
27.5	50.10	
		−1.05
30	49.05	
		1.10
35	50.15	

Three situations may occur.

1. If the data show no trend, we expect a δ^2/s^2 near 2.
2. If the data follow some curve, we expect a δ^2/s^2 less than 2.
3. If the data oscillate rapidly, we expect a δ^2/s^2 greater than 2.

Example. Five mixtures containing glucose are allowed to stand for a total of 35 weeks. At approximately 4-week intervals, the % glucose concentration is determined. The means (each being the mean of five observations) are given in Table 2.4. We wish to test at the 5% significance level whether the mean % glucose concentration increases with time.

We compute

$$\frac{\delta^2}{s^2} = \frac{\sum (\bar{x}_{i+1} - \bar{x}_i)^2/(k-1)}{\sum (\bar{x}_i - \bar{\bar{x}})^2/(k-1)}$$

$$= \frac{\sum (\bar{x}_{i+1} - \bar{x}_i)^2}{\sum (\bar{x}_i - \bar{\bar{x}})^2}$$

$$= \frac{\sum (\bar{x}_{i+1} - \bar{x}_i)^2}{\sum \bar{x}_i^2 - (\sum \bar{x}_i)^2/k}$$

$$= \frac{31.1247}{75.8454}$$

$$= 0.4104$$

In Appendix Table 14 for $k = 9$, the 95% critical value for δ^2/s^2 is 1.0244. Since $0.4104 < 1.0244$, δ^2/s^2 is significant at the 5% level. Thus, a trend in mean % glucose concentration as a function of time does exist.

BIBLIOGRAPHY

2.1. Army Ordnance Corps. Tables of the Cumulative Binomial Probabilities, by Leslie E. Simon and Frank E. Grubbs. Washington, Office of the Chief of Ordnance, 1952. (ORDP 20–1.) Available at GPO, Washington.

2.2. Dixon, W. J., and F. J. Massey, Jr. An Introduction to Statistical Analysis. New York, McGraw-Hill, 1951. Chap. 9, 13 (Sec. 5 and 6), 16, 17, and 19.

2.3. Frankford Arsenal. Statistical Manual. Methods of Making Experimental Inferences, 2nd rev. ed., by C. W. Churchman. Philadelphia, Frankford Arsenal, June 1951.

2.4. Freeman, H. Industrial Statistics. New York, Wiley, 1942. Chap. I.

2.5. National Bureau of Standards. Tables of the Binomial Probability Distribution. Washington, GPO, 1949. (Applied Mathematics Series 6.)

2.6. National Defense Research Committee. Theory of Probability With Applications, by Henry Scheffé. Washington, NDRC, February 1944. (Armor and Ordnance Report A-224 (OSRD 1918) Division 2.)

2.7. Romig, Harry G. 50–100 Binomial Tables. New York, Wiley, 1952.

2.8. Wilks, S. S. Elementary Statistical Analysis. Princeton, N. J., Princeton University Press, 1949. Chap. 10 and 11.

Chapter 3

TESTS AND CONFIDENCE INTERVALS FOR
STANDARD DEVIATIONS

3.0. Introduction

In this chapter, methods are given for (1) testing, by examination of a random sample, the hypothesis that the standard deviation σ (or the variance σ^2) of a normal population has a certain hypothetical value, specified before the experiment; (2) calculating confidence limits for the estimate of σ; and (3) comparing two or more standard deviations.

What is the importance of the population standard deviation? It is a measure of dispersion; in other words, a measure of lack of precision. As such, it includes both lack of reproducibility and whatever measurement error is present. A weapon cannot be used for tasks requiring precision if test-firing shows that the standard deviation in distance from a fixed target is large. On the other hand, if the standard deviation is small, but the average distance of shots from the target line is large, it is possible that a simple aiming correction will center the shots on the target. If the correction can be made without increasing the dispersion, the weapon will be both accurate and precise.

In production, a sudden increase in variability may indicate the appearance of a production fault, such as the maladjustment of a machine or the dulling of a cutter blade.

The tests presented here are based on the assumption that the population is normal. This restriction can often be relaxed, especially in cases involving large samples. All samples are taken to be random samples. (See Chapter 4 for tests of normality and tests of randomness.)

All the significance tests of Chapter 3 involve fixed sample sizes. However, if it is practical to experiment on one item at a time without knowing the total number that will be required, the most economical method is provided by "sequential analysis." (See Sec. 8.4 and its references.)

3.1. Computation of s

The sample variance, and best estimate of the population variance σ^2, is

$$s^2 = \frac{1}{n-1} \sum_{i=1}^{n} (x_i - \bar{x})^2 = \frac{1}{n(n-1)} \left[n \sum_{i=1}^{n} x_i^2 - \left(\sum_{i=1}^{n} x_i \right)^2 \right]$$

The latter form is preferred for computation. The sum of the x_i's and the sum of the x_i^2's can be computed simultaneously on most desk computers by accumulated multiplication.

If k samples from populations having a common variance are available for estimating the variance, the pooled estimate is

$$s_0^2 = \frac{(n_1 - 1)s_1^2 + (n_2 - 1)s_2^2 + \cdots + (n_k - 1)s_k^2}{n_1 + n_2 + \cdots + n_k - k}$$

where $(n_1 - 1)s_1^2$ is the sum of the squared deviations of observations in the first sample from their mean, and so on. If the k samples are from populations that may have different means, s_0^2 should be used instead of an over-all computed s^2, as the latter will overestimate the variance by including the differences among the means. The pooled estimate s_0 may be used to make significance tests and to find confidence intervals for the common standard deviation of the k populations with the number of degrees of freedom $n_1 + n_2 + \cdots + n_k - k$. (An example of the use of s_0^2 occurs in Sec. 2.5.3b.)

If the x values are coded by letting $y_i = mx_i + b$, where m and b are constants,

$$\bar{x} = \frac{1}{m} (\bar{y} - b) \qquad \text{and} \qquad s_x^2 = \frac{1}{m^2} s_y^2$$

Examples of the computations appear throughout this chapter; see, for instance, Sec. 3.2.2.

If means and standard deviations must be obtained from data that have been grouped together in intervals of length h, so that all observations lying in one interval are considered to be at the midpoint of the interval, we may employ "Sheppard's correction" to improve the approximation. This correction calls for subtraction of $h^2/12$ from the s^2 computed using midpoints. The correction is appropriate if the sample is taken from a distribution that tapers to zero at the ends. (See Ref. 4.4, pp. 359–63.)

If y is a linear function of several independent variables $x_1, x_2, \cdots, x_k$,

$$y = a_1x_1 + a_2x_2 + \cdots + a_kx_k + b$$

then the variance of y is

$$\sigma_y^2 = a_1^2\sigma_{x_1}^2 + a_2^2\sigma_{x_2}^2 + \cdots + a_k^2\sigma_{x_k}^2$$

The standard deviation of y may be found in terms of the standard deviations of the x's as

$$\sigma_y = (a_1^2\sigma_{x_1}^2 + a_2^2\sigma_{x_2}^2 + \cdots + a_k^2\sigma_{x_k}^2)^{\frac{1}{2}}$$

Example. If $y = (x_1 + x_2 + \cdots + x_n)/n$, where the x_i's form a random sample from a population with standard deviation σ_x, then $\sigma_y^2 = n(1/n)^2\sigma_x^2 = \sigma_x^2/n$. Hence $\sigma_{\bar{x}} = \sigma_x/\sqrt{n}$.

Example. If $y = \bar{x}_1 - \bar{x}_2$, where $\bar{x}_1$ and $\bar{x}_2$ are the means of independent random samples of size n_1 and n_2, then

$$\sigma_y^2 = \sigma_{\bar{x}_1}^2 + \sigma_{\bar{x}_2}^2 = \frac{\sigma_{x_1}^2}{n_1} + \frac{\sigma_{x_2}^2}{n_2}$$

If $y = F(x_1, x_2, \cdots, x_k)$, and the changes Δx_i in the x's are small, the differential dy approximates the change Δy in y. Thus

$$\Delta y \cong dy = \frac{\partial F}{\partial x_1}\Delta x_1 + \frac{\partial F}{\partial x_2}\Delta x_2 + \cdots + \frac{\partial F}{\partial x_k}\Delta x_k$$

If the changes Δx_i in the x's are independent, the variance in Δy may be approximated by

$$\sigma_y^2 \cong \left(\frac{\partial F}{\partial x_1}\right)^2\sigma_{x_1}^2 + \left(\frac{\partial F}{\partial x_2}\right)^2\sigma_{x_2}^2 + \cdots + \left(\frac{\partial F}{\partial x_k}\right)^2\sigma_{x_k}^2$$

with the partial derivatives evaluated at nominal values of $x_1, x_2, \cdots, x_k$.

Example. The vacuum velocity of a rocket after burning can be computed as

$$v = v_g \log_e \frac{m_b + m_p}{m_b}$$

The gas velocity v_g, the "burnt" rocket weight m_b, and the propellant weight m_p are assumed to have small independent variations. Find the variance of the vacuum velocity v in terms of the variances of v_g, m_b, and m_p.

Taking the total differential, we have

$$dv = \frac{\partial v}{\partial v_g} \Delta v_g + \frac{\partial v}{\partial m_b} \Delta m_b + \frac{\partial v}{\partial m_p} \Delta m_p$$

$$= \left(\log_e \frac{m_b + m_p}{m_b} \right) \Delta v_g + \left[\frac{-m_p v_g}{m_b(m_b + m_p)} \right] \Delta m_b + \left(\frac{v_g}{m_b + m_p} \right) \Delta m_p$$

and the variance of v is approximately

$$\sigma_v^2 \cong \left(\log_e \frac{m_b + m_p}{m_b} \right)^2 \sigma_{v_g}^2 + \left[\frac{-m_p v_g}{m_b(m_b + m_p)} \right]^2 \sigma_{m_b}^2 + \left(\frac{v_g}{m_b + m_p} \right)^2 \sigma_{m_p}^2$$

If a new type of rocket is to have $v_g = 8,000$ feet per second, $m_b = 50$ pounds, and $m_p = 50$ pounds, we can compute

$$\sigma_v^2 \cong 0.480\sigma_{v_g}^2 + 6,400\sigma_{m_b}^2 + 6,400\sigma_{m_p}^2$$

3.2. χ^2 Test for σ in a Normal Distribution

3.2.1. Assumptions

(1) The population has a normal distribution.

(2) The sample is a random sample.

3.2.2. Test

To test the null hypothesis that the normal population has standard deviation $\sigma = \sigma_0$ at the significance level α on the basis of a sample of size n, we compute the sum of the squared deviations from the sample mean, $(n - 1)s^2$, and form the chi-square statistic

$$\chi^2 = \frac{(n - 1)s^2}{\sigma_0^2}$$

Appendix Table 4 gives values of the χ^2 distribution for $f = n - 1$ degrees of freedom. If the computed χ^2 is less than $\chi^2_{1-\alpha/2,n-1}$ for $P = 1 - \alpha/2$, or greater than $\chi^2_{\alpha/2,n-1}$ for $P = \alpha/2$, we reject the null hypothesis that the standard deviation is σ_0, for if the standard deviation actually is σ_0, such a large or such a small sample variance will occur with a probability of only α. (See Sec. 1.1.7 for a general discussion of statistical tests.) The number $\chi^2_{\alpha/2,n-1}$ is the χ^2 value exceeded with probability $\alpha/2$ for $n - 1$

degrees of freedom. Similarly, for $n - 1$ degrees of freedom, $\chi^2_{1-\alpha/2, n-1}$ is the χ^2 value exceeded with probability $1 - \alpha/2$. (See Sec. 8.4.2 for a sequential test for σ.)

Example (Equal-Tails Test). The data in Table 3.1 represent the burning times of 13 grains. Assuming that the distribution of burning

TABLE 3.1. ACTUAL AND CODED BURNING
TIMES OF 13 GRAINS

Burning time, sec (x_i)	Coded value $(y_i = 1{,}000x_i - 500)$
0.516	16
0.508	8
0.517	17
0.529	29
0.501	1
0.521	21
0.539	39
0.509	9
0.521	21
0.532	32
0.547	47
0.504	4
0.525	25

times is normal, test at the 5% level of significance whether the standard deviation of the population is 0.01 (i.e., $\sigma = \sigma_0 = 0.01$). We compute

$$\sum y_i = 269$$

$$\sum y_i^2 = 7{,}809$$

$$(n - 1)s_y^2 = \frac{n \sum y_i^2 - (\sum y_i)^2}{n}$$

$$= \frac{(13)(7{,}809) - (269)^2}{13} = 2{,}242.769$$

$$(n - 1)s_x^2 = 2{,}242.769 \times 10^{-6}$$

$$\chi^2 = \frac{(n - 1)s_x^2}{\sigma_0^2} = \frac{2{,}242.769 \times 10^{-6}}{10^{-4}} = 22.43$$

From Appendix Table 4 for 12 degrees of freedom, we read

$$\chi^2_{.975,12} = 4.404 \quad \text{and} \quad \chi^2_{.025,12} = 23.34$$

Since 22.43 lies inside this interval, we cannot reject the hypothesis that $\sigma = 0.01$.

Example (One-Sided Test). If in the case of burning time we are interested only in whether σ *exceeds* 0.01, and are quite satisfied with any smaller standard deviation, we compare $\chi^2 = 22.43$ with

$$\chi^2_{.05,12} = 21.03$$

Since 22.43 exceeds 21.03, we reject at the 5% level of significance the hypothesis that $\sigma \leqslant 0.01$, and conclude that σ exceeds 0.01.

a. Determination of Sample Size. We can determine the necessary sample size for a one-sided test of the null hypothesis that the standard deviation σ equals some hypothetical value σ_0 rather than alternative values greater than σ_0 at the significance level α. (As defined in Sec. 1.1.8, the significance level α is the risk of rejecting the hypothesis that the standard deviation is σ_0 when the standard deviation actually is σ_0.) We want the risk to be not more than β of accepting the hypothesis that σ is σ_0 when actually σ has the alternative value $\sigma_1 = \lambda\sigma_0$. Appendix Chart VII gives OC curves for several sample sizes for a one-sided test $(\lambda > 1)$ conducted at the 5% level of significance. Compute

$$\lambda = \frac{\sigma_1}{\sigma_0}$$

and enter the chart with λ and β. Use the sample size indicated on the next lower curve, or interpolate between curves.

Example. We wish to make a one-sided test at the 5% level to determine whether the standard deviation of length of grains made by a new process is larger than the standard deviation obtained under the former process. We will accept a risk of 0.1 of not detecting a standard deviation half again as large as the old standard deviation. How large a sample size should be used?

In this case $\lambda = 1.5\sigma_0/\sigma_0 = 1.5$, and $\beta = 0.1$. Appendix Chart VII shows that a sample size of 28 is necessary. Using this sample size assures us that we can accept the null hypothesis $\sigma = \sigma_0$, or the alternative hypothesis $\sigma = \sigma_1 = 1.5\sigma_0$, with satisfactorily small risks of being wrong in either case.

3.2.3. Confidence Intervals

A set of $100(1 - \alpha)\%$ confidence limits for the standard deviation σ of a normal population may be computed from a sample of size n as

$$\left[\frac{(n-1)s^2}{\chi^2_{a/2,n-1}}\right]^{1/2} \quad \text{and} \quad \left[\frac{(n-1)s^2}{\chi^2_{1-a/2,n-1}}\right]^{1/2}$$

where $(n - 1)s^2 = \Sigma(x_i - \bar{x})^2$. Unlike the confidence limits for the mean (Sec. 2.1.3), which are symmetric about the estimate $\bar{x}$, these limits are not symmetric about the estimate s.

Alternatively, Appendix Table 8 enables us to obtain 90, 95, or 99% confidence limits for σ from a sample of size n by multiplying the factors b_1 and b_2, given for $f = n - 1$ degrees of freedom, by the estimate s computed from the sample.

Example. A sample of 20 gage measurements of pressure has a sample standard deviation of 11 pounds per square inch. What are 95% confidence limits for the population standard deviation σ?

From Appendix Table 4, we find

$$\chi^2_{.025,19} = 32.85 \quad \text{and} \quad \chi^2_{.975,19} = 8.907$$

With $s = 11$, we have as 95% confidence limits for σ

$$\left(\frac{19}{32.85}\right)^{1/2}(11) = 8.4 \text{ psi} \quad \text{and} \quad \left(\frac{19}{8.907}\right)^{1/2}(11) = 16.1 \text{ psi}$$

Notice that the limits are not symmetric; that is, they are not equally far from s.

The following method is preferable in practice. From Appendix Table 8, for $f = n - 1 = 19$ the factors $b_1 = 0.760$ and $b_2 = 1.461$ yield the 95% confidence limits

$$(0.760)(11) = 8.4 \text{ psi} \quad \text{and} \quad (1.461)(11) = 16.1 \text{ psi}$$

a. Determination of Sample Size. We can determine the sample size necessary to estimate σ within $p\%$ of its true value with confidence coefficient $1 - \alpha$. Appendix Chart IX gives the desired sample size as a function of p and f.

Example. In estimating the precision of a new weapon, how large a sample size should be used to specify the standard deviation within 30% of its true value with 95% confidence?

From Appendix Chart IX, for $p = 30\%$ on the curve for confidence coefficient 0.95, we find 21 degrees of freedom. Since for our test the number of degrees of freedom is $n - 1$, the necessary sample size is 22.

3.3. *F* Test for σ_1/σ_2 for Normal Distributions

3.3.1. Assumptions

(1) The populations have normal distributions.

(2) The samples are random samples drawn independently from the two populations.

3.3.2. Test

To test the null hypothesis that the ratio σ_1^2/σ_2^2 of the variances of two normal populations is 1 (i.e., $\sigma_1 = \sigma_2$) at the significance level α, on the basis of a sample of size n_1 from population 1 and an independent sample of size n_2 from population 2, compute the statistic

$$F = \frac{s_1^2}{s_2^2}$$

with the larger s^2 in the numerator so that the computed F is always greater than 1. Enter Appendix Table 5 of the F distribution with $f_1 = n_1 - 1$ degrees of freedom for the numerator and $f_2 = n_2 - 1$ degrees of freedom for the denominator. Reject the hypothesis if F exceeds the tabled value $F_{\alpha/2}(n_1 - 1, n_2 - 1)$ for $P(F) = \alpha/2$, because a ratio F that large or larger can occur with probability only α if the null hypothesis holds. For an equal-tails test, we force the computed F to be greater than 1 by the choice of numerator and denominator; hence, we use $F > F_{\alpha/2}$ as the rejection region instead of the double region $F > F_{\alpha/2}$ and $F < F_{1-\alpha/2}$.

For an equal-tails test of the null hypothesis $\sigma_1^2/\sigma_2^2 = R \neq 1$, compute F as the ratio of s_1^2 and Rs_2^2, using the larger quantity in the numerator.

A *one-sided test* is appropriate if we are interested only in whether population 1, specified before observing the data, has greater variability than population 2. We compute

$$F = \frac{s_1^2}{s_2^2}$$

where s_1^2 represents the estimated variance of population 1 (not necessarily greater than s_2^2 as in the equal-tails test). If F exceeds the tabled value $F_\alpha(n_1 - 1, n_2 - 1)$, we reject the null hypothesis that $\sigma_1 \leqslant \sigma_2$ and decide that population 1 has more variability than population 2. If the computed $F < 1$, we can immediately accept the null hypothesis that $\sigma_1 \leqslant \sigma_2$, since the tabled value of F is always greater than 1.

Figure 3.1 shows the probability density function of F for two pairs of numbers of degrees of freedom.

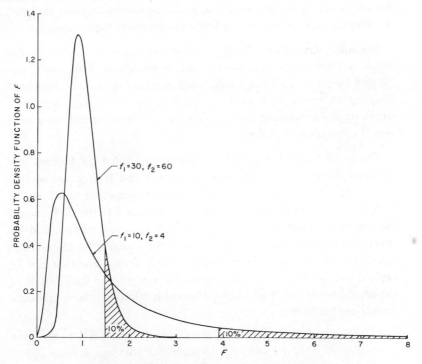

FIG. 3.1. Frequency Distribution of $F = s_1^2/s_2^2$.

Example (Equal-Tails Test). To determine whether two types of rocket have different dispersions and should, therefore, be recommended for different uses, an F test is made on data already available. From a random sample of 61 rockets of one type, the computed s^2 is 1,346.89 mils². From a random sample of 31 rockets of the other type, the computed s^2 is 2,237.29 mils². Is there a significant difference at the 5% level in the standard deviations?

Since 2,237.29 exceeds 1,346.89, we label the sample of 31 rockets Sample 1. To test the null hypothesis $\sigma_1 = \sigma_2$, we compute

$$F = \frac{s_1^2}{s_2^2} = \frac{2,237.29}{1,346.89} = 1.66$$

From Appendix Table 5, $F_{.025}(30, 60) = 1.82$. Since 1.66 is less than 1.82, we cannot reject the null hypothesis that the population standard deviations are equal. We need not, as far as this experiment indicates, differentiate between the two rockets with respect to their dispersion.

Example (One-Sided Test). A personnel department plans to distribute a form to job applicants to aid in placement. Two forms, one 2 pages long and one 3 pages long, involving grading systems, have been designed. A test is made to determine whether the 3-page form has a larger standard deviation of scores than the 2-page form. Here a wide spread of scores is preferred.

The 3- by 5-inch index cards of information filled out by each new employee during one week's hiring are shuffled, and 16 cards are selected at random. The 16 employees thus selected are requested to fill out the 3-page form (Sample 1). Their cards are returned to the deck and all the cards are reshuffled. Again 16 cards are drawn at random, and the employees thus selected are asked to fill out the 2-page form (Sample 2). Notice that the two samples are independently drawn. Since the first 16 cards drawn are replaced, their selection has no effect on the selection of the second set of 16. If an employee's card is selected in both drawings, he fills out both forms.

After scoring, the 2-page form gives $s_2^2 = 12.3$, and the 3-page form gives $s_1^2 = 40.2$. Is the standard deviation for the 3-page form significantly greater at the 5% level than that of the 2-page form?

To test the null hypothesis $\sigma_1 \leqslant \sigma_2$, we compute

$$F = \frac{40.2}{12.3} = 3.27$$

and compare with $F_{.05}(15, 15) = 2.40$. Since $F > F_{.05}$, we reject the null hypothesis, and conclude that the standard deviation of scores for the 3-page form is greater than that for the 2-page form. If the 2-page form had shown greater standard deviation (as could often happen under the null hypothesis), we would have kept to our one-sided test by merely stating that no significant departure had been shown.

a. Determination of Sample Size. Suppose we are using the one-sided test described in this section to determine at the significance level α whether σ_1 exceeds σ_2, with an assigned risk β of accepting the null hypothesis that $\sigma_1/\sigma_2 = 1$ when actually $\sigma_1/\sigma_2 = \lambda > 1$. If samples of equal size $n_1 = n_2 = n$ are to be taken from the two populations, the necessary sample size n can be found in Appendix Chart VIII opposite λ and β for the special case $\alpha = 0.05$.

Example. It is suspected that the aiming error of a pilot at high altitudes will be greater (1) without the use of a pressurized suit, than (2) with a pressurized suit. A test is to be made by having a pilot fire $n_1 = n_2 = n$ shots under each condition. We will accept a risk of 0.1 of deciding that the standard deviations σ_1 and σ_2 of the shots are equal, when actually $\sigma_1 = 2\sigma_2$. What sample size should be used if the test is to be made at the 5% level of significance?

From Appendix Chart VIII for $\lambda = 2$ and $\beta = 0.1$, we find $n = 20$.

3.3.3. Confidence Intervals

A set of $100(1 - \alpha)\%$ confidence limits for the ratio σ_1/σ_2 may be computed. The lower limit is

$$\frac{s_1}{s_2} \frac{1}{\sqrt{F_{\alpha/2}(n_1 - 1,\ n_2 - 1)}}$$

and the upper limit is

$$\frac{s_1}{s_2} \frac{1}{\sqrt{F_{1-\alpha/2}(n_1 - 1,\ n_2 - 1)}} = \frac{s_1}{s_2} \sqrt{F_{\alpha/2}(n_2 - 1,\ n_1 - 1)}$$

The latter form of the upper limit should be used for computation.

a. Determination of Sample Size. We can determine the sample size necessary for estimating the ratio σ_1/σ_2 within $p\%$ of its true value with confidence coefficient 0.90, 0.95, or 0.99 from the number of degrees of freedom $f_1 = f_2 = f$, shown in Appendix Chart X.

Example. How large a sample should be drawn from each of two populations in order to estimate the ratio σ_1/σ_2 within 20% of its true value with confidence coefficient 0.95?

From Appendix Chart X for $p = 20\%$, we find the number of degrees of freedom $f = 99$. Therefore the sample size $n_1 = n_2 = n = 99 + 1 = 100$.

NOTE: From Appendix Charts IX and X it follows that a *total* of slightly more than *four* times as many observations are needed to compare two unknown standard deviations with any given accuracy as are needed to compare one unknown standard deviation with a known value with the same accuracy.

3.4. M Test for Homogeneity of Variances

3.4.1. Assumptions

(1) The populations have normal distributions.

(2) The samples are random samples drawn independently from the respective populations.

3.4.2. Test

To test the null hypothesis that k populations, from which k samples of sizes $n_1, n_2, \cdots, n_k$ have been drawn, have the same variances, we compute

$$M = \phi \log_e \frac{\sum\limits_{i=1}^{k} f_i s_i^2}{\phi} - \sum\limits_{i=1}^{k} f_i \log_e s_i^2$$

where

$f_i = n_i - 1$, the number of degrees of freedom in computing s_i^2 for the ith sample

$$\phi = \sum_{i=1}^{k} f_i$$

s_i^2 = estimate of the variance of the ith population

$\log_e = 2.3026 \log_{10}$

If we wish to use common logarithms (base 10), we substitute "$\log_{10}$" in the formula wherever "$\log_e$" occurs and multiply the final result by 2.3026.

Appendix Table 6 gives values of M at the 5% significance level. If the computed value of M exceeds all the values in the rows for the given k, it is significant at the 5% level, and we reject the null hypothesis that the variances are all equal. If M is less than all the values in the rows, it is not significant at the 5% level, and we cannot reject the null hypothesis. If M falls within the range of tabled values for the given k, a more refined technique is necessary. We compute

$$c_1 = \sum_{i=1}^{k} \frac{1}{f_i} - \frac{1}{\phi}$$

To be significant, M must exceed the appropriate value under c_1 (interpolating between values of c_1 if necessary), opposite the given k. For the computed c_1, the appropriate critical value lies between the value in row (a) and the value in row (b).

The (a) value is a maximum, obtained when all the values f_i are equal. The (b) value is a minimum, obtained when some of the values $f_i = 1$ and the rest are infinite. In case neither condition (a) nor condition (b) applies and the computed M lies between the (a) and (b) values (as will rarely happen), this borderline case may be attacked by methods of Ref. 3.5. Ref. 3.5 gives both 1 and 5% points.

Refer to Chapter 5 for a test of homogeneity of means, which can be applied if the M test shows that the variances are homogeneous.

Example. Table 3.2 gives statistics on 10 samples of 6 from the production of 10 different machines. Is the variability the same for all machines?

TABLE 3.2. DATA ON SAMPLES FROM 10 MACHINES

Machine	s_i^2	$\log_e s_i^2$	$\bar{x}_i$
1	95.46	4.559	11.7
2	26.16	3.264	6.2
3	9.20	2.219	1.0
4	14.70	2.688	2.5
5	14.96	2.705	3.2
6	14.16	2.650	3.2
7	25.36	3.233	6.2
8	24.16	3.185	8.8
9	3.06	1.118	6.3
10	26.70	3.285	4.5
Total	253.92	28.906	53.6

With $k = 10$, $n_i = n = 6$, and $f_i = f = 5$, we compute

$$\phi = \sum_{i=1}^{k} f_i = 50$$

$$\frac{\sum f_i s_i^2}{\phi} = \frac{1}{10} \sum s_i^2 = 25.392$$

$$\log_e \frac{\sum f_i s_i^2}{\phi} = 3.234$$

$$M = (50)(3.234) - (5)(28.906) = 17.17$$

In Table 6 opposite $k = 10$, we see that the values range from 16.92 to 19.89. Since this range includes our M value, we go on to compute

$$c_1 = \sum \frac{1}{f_i} - \frac{1}{\phi} = 2 - 0.02 = 1.98$$

The value $M = 17.17$ is less than the (a) values for both $c_1 = 1.5$ and $c_1 = 2.0$; hence no interpolation is necessary for $c_1 = 1.98$. We conclude that no difference among the variances has been shown at the 5% significance level. (Note that here the (a) value is the pertinent one, since all the f_i are equal.)

BIBLIOGRAPHY

3.1. Dixon, W. J., and F. J. Massey, Jr. An Introduction to Statistical Analysis. New York, McGraw-Hill, 1951. Chap. 8 and 16.

3.2. Eisenhart, C. "Planning and Interpreting Experiments for Comparing Two Standard Deviations," in Selected Techniques of Statistical Analysis, ed. by C. Eisenhart, M. W. Hastay, and W. A. Wallis. New York, McGraw-Hill, 1947. Chap. 8.

3.3 Frankford Arsenal. Statistical Manual. Methods of Making Experimental Inferences, 2nd rev. ed., by C. W. Churchman. Philadelphia, Frankford Arsenal, June 1951.

3.4. Hoel, P. G. Introduction to Mathematical Statistics, 2nd ed. New York, Wiley, 1954. Chap. 11.

3.5. Thompson, Catherine M., and Maxine Merrington. "Tables for Testing the Homogeneity of a Set of Estimated Variances," BIOMETRIKA, Vol. 33 (1943–46), pp. 296–304.

Chapter 4

TESTS OF DISTRIBUTIONS AS A WHOLE AND ALLIED PROBLEMS

4.0. Introduction

In Chapter 2, tests were given to determine whether a sample came from a population having mean μ. In Chapter 3, there were tests designed to show whether a sample came from a normal population having standard deviation σ. In this chapter, tests are given to determine whether a sample comes from a population having a distribution of any specified form. Such information is often valuable. For instance, several of the tests of Chapter 2 and Chapter 3 required a normal distribution. With the criteria of this chapter, we can test whether a normal-distribution requirement is satisfied.

Several miscellaneous topics are treated, such as randomness, sensitivity testing, independence, gross errors, and tolerance intervals.

4.1. Run Test for Randomness

In all the tests given in this Manual, it is required that the sample be drawn at *random* from its population. (See Sec. 1.1.1.) Sometimes it is difficult, without examining the sample itself, to be sure that the selection does not involve some hidden trend. The run test for randomness, which can be applied as a one-sided or an equal-tails test, is illustrated by the following example.

Example. Coded measurements of the weights of 20 grains are recorded in Table 4.1 in the order in which they were obtained. Test at the 5% level of significance whether the 20 weights represent a random sample from which to estimate the average weight for the population of all such grains.

We find that the median of the 20 readings is 26 (that is, the average of the tenth and eleventh when the readings are arranged in the order of

83

TABLE 4.1. CODED MEASUREMENTS OF
WEIGHTS OF 20 GRAINS

12	23	29	28
14	25	10	28
22	27	25	30
21	25	30	30
22	28	30	28

size). We mark with minus signs readings less than 26 and with plus signs those greater than 26, as in Table 4.2. Here we are interested not just in the numbers of these signs (as in the sign test, Sec. 2.5.2a), but in how they are interspersed. If the sample is random, we should expect low values and high values (values below 26 and values above 26) to be fairly well scattered; for instance, we should not expect all the low values to precede all the high values. On the other hand, we should not expect them to alternate. If we count an unbroken sequence of plus signs or minus signs as one "run," the data show that the number of runs v is 6.

Appendix Table 10 is a table of critical values for runs. The table is entered with quantities n_1 and n_2, the numbers of objects of the two kinds in the sequence. In this case $n_1 = n_2 = 10$, since there are 10 items below and 10 items above the median. Thus we find $v_{.975} = 6$ and $v_{.025} = 16$. A number of runs greater than 6 but less than 16 is consistent with the null hypothesis of randomness at the 5% level of signif-

TABLE 4.2. CODED WEIGHTS OF 20 GRAINS SEPARATED IN RUNS

			Run no.		
1	2	3	4	5	6
12 −	27 +	25 −	28 +	10 −	30 +
14 −			29 +	25 −	30 +
22 −					28 +
21 −					28 +
22 −					30 +
23 −					30 +
25 −					28 +

icance. A value of 6 or below, or of 16 or above, indicates that the hypothesis should be rejected at this level, since if the hypothesis is true, a number of runs either so small or so large would happen with probability not more than 0.05. (See Sec. 1.1.7 for a general discussion of tests of hypotheses.)

Since here the conclusion is that the sample is not random, we should look for possible causes. If the weight readings were taken on the 20 grains at the time of production and in the order of their extrusion, we should look for a factor (such as the warming-up of a machine) that would cause an increase, with time, in the weight of the grains. If rejection had been on the grounds that there were too many runs ($\geqslant 16$), we should look for a factor in production that tends to cause alternate heavy and light grains.

Another application of the run test appears in Sec. 4.7.2. For a test of trend in means, see Sec. 2.8. For a test of the randomness of occurrence of defects in production, see the example in Sec. 4.2.

4.2. The χ^2 Test for Goodness of Fit

Suppose the sample values from an experiment fall into r categories. To decide at the significance level α whether the data constitute a sample from a population with distribution function $f(x)$, we first compute the expected number of observations that would fall in each category as predicted by $f(x)$. The grouping should be arranged so that this theoretical frequency is at least five for each category. If necessary, several of the original groups may be combined in order to assure this. To compare the observed frequencies, n_i for the ith category, with the expected (theoretical) frequencies, e_i, compute

$$\chi^2 = \sum_{i=1}^{r} \frac{(n_i - e_i)^2}{e_i}$$

If the calculated value of χ^2 exceeds the value $\chi^2_{\alpha, r-1-g}$ for $f = r - 1 - g$ degrees of freedom in Appendix Table 4, reject at the significance level α the null hypothesis that the distribution function is $f(x)$. An explanation of the subtracted g follows.

If, before the experiment, the distribution function $f(x)$ to be tested is completely specified (for instance, in an experiment to determine whether a coin is "true" by comparing the numbers of heads and tails tossed with those predicted by the binomial distribution with proportion $p = \frac{1}{2}$), then $g = 0$. If, before the experiment, we can specify the type or form of the distribution $f(x)$ which we want to test, but we cannot specify $f(x)$ completely (for instance, in an experiment to see whether aiming errors have a normal distribution, or in an experiment to see whether burning time has a normal distribution with standard deviation 0.01 second), then g stands for the number of quantities necessary to complete the specification. These quantities must be obtained as estimates from the experimental data themselves. For instance, to decide whether aiming errors have a normal distribution, we compute $\bar{x}$ from the sample as an estimate of μ and s as an estimate of σ, and since we use $g = 2$, the test shows whether the observations are consistent with the hypothesis that the data come from a normal distribution. To decide whether burning time has a normal distribution with standard deviation 0.01 second, we need to estimate μ only; here $g = 1$.

Example. In Sec. 1.2.3, plant records are given showing the frequency of occurrence of x defects in a production day; these data are repeated in Table 4.3. Test at the 5% level of significance whether defects are occurring randomly.

TABLE 4.3. FITTING A POISSON DISTRIBUTION

No. of defects produced in a day, x	Actual no. of occurrences of x defects in a day, n_i	Theoretical no. of occurrences of x defects in a day, e_i
0	102	95.3
1	59	71.1
2	31	26.5
3 ⎫	8 ⎫	6.6 ⎫
4 ⎬	0 ⎬ 9	1.2 ⎬ 8.0
5 ⎥	1 ⎥	0.2 ⎥
6 or more ⎭	0 ⎭	0.0 ⎭

If defects are occurring randomly, the relative frequencies should follow the Poisson distribution

$$P(x) = \frac{m^x e^{-m}}{x!} \qquad (x = 0, 1, 2, \cdots)$$

Since we do not know which specific Poisson distribution to expect (namely, which value of m to use), we must use for m the mean of the sample, 0.74627. We multiply the total number of days, 201, by the theoretical relative frequency in each category, $P(x)$, to obtain the theoretical number in each category.

In order to have the theoretical frequency in each category at least 5, we group the last four figures together. We then compute

$$\chi^2 = \sum_{i=1}^{4} \frac{(n_i - e_i)^2}{e_i}$$

$$= \frac{(6.7)^2}{95.3} + \frac{(-12.1)^2}{71.1} + \frac{(4.5)^2}{26.5} + \frac{(1.0)^2}{8.0} = 3.42$$

Taking $g = 1$, because we estimated m from the sample, we find from Appendix Table 4 that $\chi^2_{.05,4-1-1} = 5.991$. Since 3.42 does not exceed this value, we have no reason to reject the hypothesis that defects are occurring randomly.

4.3. Testing the Fit of a Normal Distribution

To test the hypothesis that a random sample comes from a population having a normal distribution, we may fit a normal curve to the data and then test to see whether the hypothesis is justified. The testing may be done by means of a χ^2 test for goodness of fit, in which case the mean and standard deviation for the fitted normal curve should be estimated from the grouped sample data. (See Chernoff and Lehmann, *Annals of Mathematical Statistics,* vol. 25, pp. 579–586 for the error in applying the χ^2 test when ungrouped data are used in fitting.)

An obvious but rough check on normality can be made by plotting the fitted normal curve to the same scale as the histogram of the grouped data. (See Fig. 4.1 below.) A more convenient graphical approach calls for the use of normal-probability graph paper (Keuffel & Esser Co., No. 358–23 and 359–23). If the population is normal, a plot of the sample cumulative percentage frequencies should approximate a straight line. However, in both of these graphical approaches there will be difficulty in judging by eye how much departure from the ideal pattern we must expect. Even the plot for a sample of 1,000 drawn from a population known to be normal often exhibits substantial irregularity.

Besides the approach through goodness of fit of the best possible normal curve, there are other tests such as the b_1, b_2, and a tests; however, many statisticians now consider the χ^2 test superior.

Example. A grouped sample of 300 observations of lateral deflection yields $\bar{x} = 0.88$ mil and $s = 52.53$ mils as estimates of the population mean and standard deviation. The computation was made from the grouped data of Table 4.4, using Sheppard's correction. Test at the 5% level of significance whether these data came from a normal distribution.

The observations were to the nearest 0.1 mil; therefore, intervals beginning at -150.05, -130.05, etc., were used to avoid ambiguity in grouping. Each group end-point is shown in column 2 as a number of s units away from $\bar{x}$, so that Appendix Table 1 (the cumulative normal distribution table) can be used to find column 3. Differencing column 3 gives the relative frequency for each group as predicted by the normal distribution. This frequency, shown in column 4, is multiplied by 300 to give the theoretical frequencies e_i of column 5. Since the theoretical frequencies are very small in the first and last few groups, all deflections less than -110.05 are combined into one new group and all deflections greater than 109.95 into another. Because the net number of groups is 13, and sample statistics, $\bar{x}$ and s, were used to estimate the corresponding population measures, μ and σ, the number of degrees of freedom for χ^2 is $13 - 1 - 2 = 10$. From Appendix Table 4, we find that $\chi^2_{.05,10} = 18.31$, and summing the values in column 7, we find that $\chi^2 = 4.89$. We conclude that at the 5% level of significance the sample distribution is consistent with the hypothesis that the parent distribution is normal.

A frequency histogram for the grouped data of Table 4.4 is given in Fig. 4.1, along with the normal curve with the same mean and standard deviation.

4.3.1. Transformations To Obtain Normality

Many physical situations produce data that are normally distributed; others, data that follow some other known distribution; and still others, data that can be transformed to normal data. For example, when the reaction rate in a chemistry experiment is proportional to the concentration of reacting substances the distribution of rates is not likely to be normal, but the

TABLE 4.4. GROUPED DATA USED IN COMPUTATION FOR χ^2 TEST OF NORMALITY

x	$\dfrac{x-0.88}{52.53}$	Cumulative normal probability	Theoretical relative frequency	Theoretical frequency, e_i	Observed frequency, n_i	$\dfrac{(n_i - e_i)^2}{e_i}$
(1)	(2)	(3)	(4)	(5)	(6)	(7)
$-\infty$	$-\infty$	0				
			0.0021	0.63 ⎫	0 ⎫	
-150.05	-2.873	0.0021				
			0.0043	1.29 ⎬ 5.19	2 ⎬ 6	0.126
-130.05	-2.492	0.0064				
			0.0109	3.27 ⎭	4 ⎭	
-110.05	-2.112	0.0173				
			0.0244	7.32	6	0.238
-90.05	-1.731	0.0417				
			0.0468	14.04	17	0.624
-70.05	-1.350	0.0885				
			0.0775	23.25	18	1.185
-50.05	-0.970	0.1660				
			0.1119	33.57	31	0.197
-30.05	-0.589	0.2779				
			0.1397	41.91	49	1.199
-10.05	-0.208	0.4176				
			0.1511	45.33	46	0.010
9.95	0.173	0.5687				
			0.1411	42.33	41	0.042
29.95	0.553	0.7098				
			0.1150	34.50	33	0.065
49.95	0.934	0.8248				
			0.0810	24.30	28	0.563
69.95	1.315	0.9058				
			0.0492	14.76	13	0.210
89.95	1.696	0.9550				
			0.0260	7.80	6	0.415
109.95	2.076	0.9810				
			0.0120	3.60 ⎫	4 ⎫	
129.95	2.457	0.9930				
			0.0047	1.41 ⎪	0 ⎪	
149.95	2.838	0.9977				
			0.0017	0.51 ⎬ 5.70	0 ⎬ 6	0.016
169.95	3.219	0.9994				
			0.0004	0.12 ⎪	2 ⎪	
189.95	3.599	0.9998				
			0.0002	0.06 ⎭	0 ⎭	
∞	∞	1				
			1.0000	300.00	300	4.890

distribution of their *logarithms* may be theoretically normal. The most common transformations are $y = \log x$ (frequently used in sensitivity testing), and $y = \sqrt{x}$. An easy way to decide whether one of these transformations is

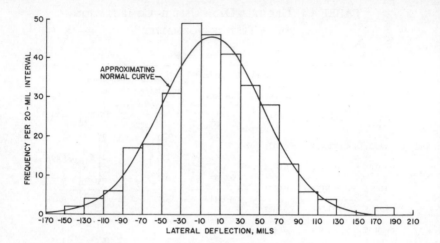

FIG. 4.1. Frequencies of Lateral Deflections, Showing the Approximating Normal Curve With $\bar{x} = 0.88$ Mil and $s = 52.53$ Mils for Sample Size 300. (Precise intervals begin at -150.05, -130.05, etc.)

likely to produce normality is to make use of special graph papers that are commercially available. For instance, if $y = \log x$ produces normality, the sample cumulative distribution curve plotted on log-probability graph paper will approximate a straight line.

4.4. Confidence Band for Cumulative Distribution

To construct, on the basis of a random sample of size n, a $100(1 - \alpha)\%$ confidence band for the cumulative distribution of the population, draw the cumulative-percentage histogram for the sample. Find in Appendix Table 11 a quantity d_α corresponding to the sample size n and the confidence coefficient $1 - \alpha$. Draw two "staircase" lines parallel to the sample histogram, one $100d_\alpha$ percentage units above the histogram, the other $100d_\alpha$ percentage units below. In the long run, confidence bands so constructed for random samples from populations will, in $100(1 - \alpha)\%$ of the cases, completely contain the population cumulative-distribution curve.

Example. In Sec. 2.5.3b, data were given on the lengths in mm of parts produced by two machines. The lengths for Machine 2, x_{2j}, arranged in order of size, are:

93.75	94.36	94.73	95.02	95.31
93.90	94.50	94.82	95.21	96.16

Construct the sample cumulative histogram and a 95% confidence band.

The cumulative histogram of these data (Fig. 4.2) shows that 0% of the sample had lengths less than 93.75; 10%, lengths less than 93.90; 20%, lengths less than 94.36; . . . ; 90%, lengths less than 96.16; and that all 10 lengths were 96.16 or less. Reference to Appendix Table 11 gives for $n = 10$ and $\alpha = 0.05$ the value $d_\alpha = 0.41$. Thus we construct the boundaries of the confidence band 41 percentage units above and 41 percentage units below the histogram.

If we superimpose on a graph like Fig. 4.2 the cumulative distribution curve $F(x)$ for some theoretical distribution $f(x)$ and if the curve leaves the confidence band at any point, we can reject at the significance level α the null hypothesis that the population was distributed according to $f(x)$. Thus, the confidence band gives us a possible test method, though the method of Sec. 4.2 is more powerful.

If we wish to test whether a distribution is of some general form, such as normal, but not to test specific values of constants (parameters), estimates of the latter are made from the sample. The test may still be made, but it lacks discrimination, in the sense that it will not reject the distribution unless its form is very different from that of the null hypothesis. It may be used correctly to reject such a distribution at a significance level no greater than α, but it will reject much too seldom.

4.5. Sensitivity Testing

"Sensitivity testing" is testing in which an increasing percentage of items fail, explode, or die as the severity of the test is increased. In such testing we cannot measure the precise severity of test (that is, the precise magnitude of the variable concerned) which would barely result in failure, but can only observe whether an applied severity results in failure, or does not. For instance, in a test to discover the minimum range to target at which a particular rocket will function as intended, we might find a given range to be either too short for proper functioning or long enough for proper functioning (but probably not the *minimum* length). The rocket expended in the first trial cannot be used again. The same difficulty arises in tests for determining the lethal dosage of a poison (Ref. 4.2). Any given dosage

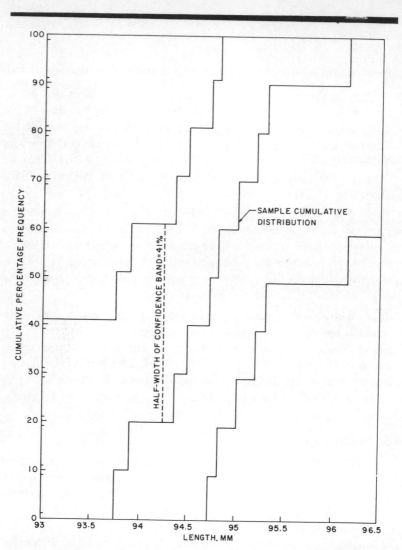

FIG. 4.2. Cumulative Histogram With 95% Confidence Band.

either kills or does not kill the experimental animal. The animal, having been affected or killed by the first trial, cannot be used again. In the determination of critical height (the minimum height from which a weight must be dropped on a charge to explode it), a charge is either demolished or changed in its characteristics as a result of the trial. In all these cases we see that each item has its own critical level of severity, so that in a whole population of items there is a probability distribution of critical levels.

Sensitivity tests for finding the mean, standard deviation, or percentiles in such cases may be analyzed by "staircase" methods (see Sec. 4.5.1 for a particular staircase method and Ref. 4.3 for others) or by "probit analysis" (see Ref. 4.2 and 4.6). Both methods depend upon an assumption of normality in the quantity being studied. A rough but quick method for the critical-height test, for instance, is to plot the percentage of items exploding below each given height on the commercially available arithmetic-probability graph paper and fit a straight line by eye. This fitting provides a check of normality also, and may suggest a transformation of the independent variable which will give approximate normality if the data do not lie near a straight line. Probit analysis does the fitting more precisely and gives a measure of the precision, but staircase methods are more efficient when they are applicable. The staircase methods are recommended when the effect on each item is known immediately or soon after each application, and when the independent variable can be readily adjusted. For instance, staircase methods are appropriate for the charge critical-height test described above. Probit analysis is recommended when it is not practical to measure one item at a time; for example, when the ultimate effects of given amounts of radiation are to be measured on a group of small animals, such as a pen of hamsters or mice. If it is inconvenient to adjust the independent variable, as in the case of tests of rockets at different temperatures, probit analysis is appropriate. For detailed application of probit analysis and the method of Sec. 4.5.1 to fuze safety and reliability, see Ref. 4.11.

4.5.1. Up-and-Down Method

A valuable and frequently used method of sensitivity testing of the staircase type is the "up-and-down," or "Bruceton," method (Ref. 4.5, pp. 279–87). Normal distribution and ready adjustment of the independent variable to any prescribed level are required. The method is especially applicable to estimating the mean μ (50% point) of the distribution because it concentrates the observations in the neighborhood of the mean; it can also be used for estimating the standard deviation σ and other percentage points, though less accurately.

Example. For estimating the distribution of the minimum functioning range of a type of rocket against a given type of target, the target would first be placed at a range y' (the best guess for μ, the range at which 50% of the rockets would function, though it is not necessary that y' be a good guess). A rocket is fired at the target. If it functions, a second rocket is fired at a target at range $y' - d$, where d is chosen beforehand as about 0.5σ if μ is to be estimated with maximum precision, and as about 1.5σ if the 10 or 90% point is to be estimated with maximum precision. If the first rocket does not function, the second rocket is fired at a target at range $y' + d$. Subsequent firings are at similarly altered ranges; i.e., the target or the launcher is moved closer by d after functioning, farther by the same amount after nonfunctioning, until all the rockets of the prescribed sample are expended. The results can be recorded as shown in Fig. 4.3, where x denotes a functioning rocket and o a nonfunctioning rocket. The notation x and o could be reversed or replaced; in some other type of test the probability of explosion might decrease as distance increases, for example.

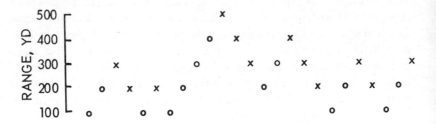

FIG. 4.3. Type of Chart Used in Up-and-Down Method of Sensitivity Testing.

The analysis involves only the symbol occurring less frequently. Since there are 13 o's and 12 x's, the latter are used. Let y_0 be the *smallest range (lowest level)* of the symbol occurring less frequently; let $y_i = y_0 + id$; and let n_i be the number of these symbols at the level y_i. Then we can summarize the data and analysis as in Table 4.5, and make the following estimates in terms of the column sums N, A, and B.

TABLE 4.5. EXAMPLE OF DATA ANALYSIS FOR UP-AND-DOWN
METHOD OF SENSITIVITY TESTING

Range	i	n_i	in_i	$i^2 n_i$
500	3	1	3	9
400	2	2	4	8
300	1	5	5	5
200	0	4	0	0
Total	.....	$N = 12$	$A = 12$	$B = 22$

Estimate of μ: $\bar{y} = y_0 + d\left(\dfrac{A}{N} \pm \dfrac{1}{2}\right)$

(+ if based on symbol at smallest value, here o;
− if based on symbol at largest value, here x)

Estimate of σ: $s = 1.620d\left(\dfrac{NB - A^2}{N^2} + 0.029\right)$

(valid only if $(NB - A^2)/N^2 > 0.3$)

Estimate of $100p\%$ point $\mu + z_p\sigma$: $\bar{y} + z_p s$

(z_p is the z for $F(z) = p$ in Appendix Table 2)

Estimate of standard deviation of $\bar{y}$: $s_{\bar{y}} = \dfrac{6s + d}{7\sqrt{N}}$

(valid only if $d < 3\sigma$)

Estimate of standard deviation of s: $s_s = \dfrac{1.1s + 0.3s^2/d}{\sqrt{N}}$

(valid only if $d < 2\sigma$)

Estimate of standard deviation of $\bar{y} + z_p s$: $s_{\bar{y}+z_p s} = \sqrt{s_{\bar{y}}^2 + z_p^2 s_s^2}$

Approximate $100(1 - \alpha)\%$ confidence limits for any parameter are obtained by taking its estimate and adding and subtracting $t_{\alpha/2, N-1}$ times the estimate of its standard deviation. Thus, applying the formulas given above to the data of Table 4.5, we have:

$$y = 200 + 100 \left(\frac{12}{12} - \frac{1}{2} \right) = 250$$

$$s = 162 \left(\frac{120}{144} + 0.029 \right) = 140$$

$$10\% \text{ point} = 250 + (-1.28)(140) = 71$$

$$s_y = \frac{6(140) + 100}{7(3.46)} = 38.8$$

$$s_x = \frac{1.1(140) + 0.3(196)}{3.46} = 61.5$$

$$s_{y-1.28x} = \sqrt{(38.8)^2 + (1.28)^2 (61.5)^2} = 87.8$$

Approximate 95% confidence limits for the 10% point are

$$71 \pm (2.201)(87.8) = 71 \pm 193 \qquad \text{or} \qquad -122 \text{ and } 264$$

In this example, the distribution is certainly confined to positive values, so -122 can be replaced by 0.

The general formulas shown above can also be used to determine sample size. In the present example, since the parameters are seen to be but poorly estimated by this sample of size 25, a larger sample size would be desirable. If an estimate of σ can be made from previous experience or design aims before the experiment, this estimate should be substituted for s in the formula shown above for $s_{\bar{y}}$, s_s, or $s_{\bar{y}+z_p x}$ in order to determine the sample size required for a prescribed precision. The total sample size is at least *twice* the N of the formulas given above. Thus, to determine the 10% point within ± 50 with 95% confidence, $(t_{.025, N-1}) s_{\bar{y}-1.28x}$ must be not greater than 50, so that $s_{\bar{y}-1.28x}$ must be about 25 or its square about 625. If, from previous experience, σ is thought to be about 100, and d is consequently taken as $1.5\sigma = 150$, then

$$625 = \frac{1}{N} \left[\left(\frac{750}{7} \right)^2 + (1.28)^2 \left(110 + \frac{3,000}{150} \right)^2 \right]$$

$$= \frac{39,100}{N}$$

$$N = 63$$

Therefore, the required number of rockets is about 130. In this way the cost of the test required for the prescribed precision may be found. If this cost is prohibitive, it becomes necessary to re-examine the importance attached to knowing the desired parameter with the prescribed precision, and to rebalance the costs and the risks involved.

4.6. Test for Independence in Contingency Tables

Table 4.6 shows a sample of size n classified according to two characteristics (I and II). The number of sample members in the cell in the ith row and the jth column is n_{ij}. Such an array is called a **contingency table.**

TABLE 4.6. CONTINGENCY TABLE FOR TWO
CHARACTERISTICS OF A SAMPLE

Character-istic II	Characteristic I				Total
	1	2	. . .	s	
1	n_{11}	n_{12}	. . .	$n_{1\bullet}$	$n_1.$
2	n_{21}	n_{22}	. . .	$n_{2\bullet}$	$n_2.$
.	.	.	.	.	.
.	.	.	.	.	.
.	.	.	.	.	.
r	n_{r1}	n_{r2}	. . .	$n_{r\bullet}$	$n_r.$
Total	$n._1$	$n._2$	. . .	$n._\bullet$	n

If the two characteristics are independent of each other, the *expected* number of sample members in any cell is calculated by simple proportion from the marginal totals to be $n_{i.}n_{.j}/n$. We use the χ^2 test to test the null hypothesis that the two characteristics are independent by comparing

$$\chi^2 = \sum_{i,j} \frac{(n_{ij} - n_{i.}n_{.j}/n)^2}{n_{i.}n_{.j}/n} = n\left(\sum_{i,j}\frac{n_{ij}^2}{n_{i.}n_{.j}} - 1\right)$$

with the value in Appendix Table 4, rejecting the hypothesis at the significance level α if $\chi^2 > \chi^2_{a,(r-1)(s-1)}$. The number of degrees of freedom, f, is $(r-1)(s-1)$. As in Sec. 4.2, the application of the χ^2 test is reliable if every *expected* frequency is at least 5.

In the contingency table, the headings of the rows and columns need not be numerical groupings. The following combinations, for instance, are possible:

Characteristic I, degree of damage: no damage, damaged slightly, completely demolished.
Characteristic II, age of structure: old, new.

Characteristic I, estimate of applicant's ability: excellent, good, fair, poor.
Characteristic II, person making estimate: Appleby, Badger, Christofer.

Example. An experiment on 75 rockets yields data on the characteristics of lateral deflection and range as shown in Table 4.7. Test at the 5% level of significance the hypothesis that these two characteristics are independent.

TABLE 4.7. CONTINGENCY TABLE FOR EXPERIMENT
WITH 75 ROCKETS

Range, yd	Lateral deflection, mils			Total
	−250 to −51	−50 to 49	50 to 199	
0 to 1,199	5	9	7	21
1,200 to 1,799	7	3	9	19
1,800 to 2,699	8	21	6	35
Total	20	33	22	75

We note that the minimum expected frequency is at least 5 and compute

$$\chi^2 = n\left(\sum_{i,j} \frac{n_{ij}^2}{n_i . n_{.j}} - 1\right)$$

$$= 75\left[\frac{5^2}{(21)(20)} + \frac{9^2}{(21)(33)} + \frac{7^2}{(21)(22)} + \frac{7^2}{(19)(20)} + \frac{3^2}{(19)(33)}\right.$$

$$\left. + \frac{9^2}{(19)(22)} + \frac{8^2}{(35)(20)} + \frac{21^2}{(35)(33)} + \frac{6^2}{(35)(22)} - 1\right]$$

$$= 75(0.13955) = 10.466$$

When the form

$$\chi^2 = n \left(\sum_{i,j} \frac{n_{ij}^2}{n_{i.}n_{.j}} - 1 \right)$$

is used, we must carry extra decimal places to obtain an accurate value, because of the order of operations (a small difference between relatively large numbers, multiplied by a large number).

From Appendix Table 4, we find that $\chi^2_{.05,(2)(2)} = 9.488$. Since 10.466 exceeds this value, we reject the hypothesis that lateral deflection and range are independent characteristics.

The χ^2 test is an approximation which improves as the cell frequencies increase. In the case of 2-by-2 tables (i.e., contingency tables for which $r = s = 2$), the approximation can be improved by applying **Yates's continuity correction.** If we have the entries shown in Table 4.8, then on the null hypothesis that the two characteristics are independent, the expected frequency (with the given marginal totals) corresponding to the observed frequency a is $(a + c)(a + b)/n$. Applying Yates's correction, we replace a by $a + \frac{1}{2}$ if $(a + c)(a + b)/n > a$, and by $a - \frac{1}{2}$ if the reverse inequality is true; then the other frequencies b, c, and d must be changed by $\frac{1}{2}$ so as to preserve the same marginal totals. (Actually, it makes no difference whether we start with a, b, c, or d.) The correction thus decreases the difference between each pair of observed and expected frequencies by $\frac{1}{2}$. If each observed frequency is so close to the expected frequency that the correction reverses the algebraic sign of the difference, then the agreement is as good as possible, the null hypothesis is immediately accepted, and no detailed calculation is necessary to test it. Otherwise Yates's correction

TABLE 4.8. CONVENTIONAL NOTATION FOR A 2-BY-2
CONTINGENCY TABLE

Character- istic II	Characteristic I		Total
	1	2	
1	a	b	$a + b$
2	c	d	$c + d$
Total	$a + c$	$b + d$	n

should always be made in 2-by-2 tables; this is especially important for small expected frequencies. If small expected frequencies, say less than 5, occur in a 2-by-2 table, it may be desirable to apply an exact test. (See Ref. 4.7, 4.8, 4.9, and 4.10. Table 38 of Ref. 4.9 gives the exact probability of each 2-by-2 table with row (or column) sums up to 15; Ref. 4.8 extends this table to row (or column) sums up to 20. See Sec. 4.7.1 for an example of the use of Yates's correction.)

For a 2-by-2 table, after the application of Yates's correction, the formula simplifies to

$$\chi^2 = \frac{n(a'd' - b'c')^2}{(a + b)(a + c)(c + d)(b + d)}$$

where the primes indicate that Yates's correction has been applied.

4.7. Consistency of Several Samples

The χ^2 test of Sec. 4.7.1 may be used on data in the form of frequencies to determine whether several samples are "consistent," that is, drawn from populations having the same distribution. For data not in the form of frequencies, the run test of Sec. 4.7.2 is applicable when only two samples are involved.

4.7.1. The χ^2 Test

Let s different samples be drawn and the degree of a certain characteristic be observed for each one, the frequency n_{ij} with which the ith degree of the characteristic $(i = 1, 2, \cdots, r)$ occurs in the jth sample being recorded in a table of the same form as Table 4.6.

To test, at the significance level α, the null hypothesis that the s samples were drawn from populations having the same distribution, proceed with the χ^2 test exactly as in Sec. 4.6, even though the statistical model is now different; that is, the column totals are not random but are the specified sample sizes.

Example. The results of testing fuzes at -40 and $130°F$ are shown in Table 4.9. Is there a significant difference at the 5% level in the proportion of failures at the two temperatures?

We apply Yates's continuity correction described in Sec. 4.6. Here the *expected* frequency of failures at $130°F$ is $(59 \times 15)/132 = 6.705 > 6$.

TABLE 4.9. RESULTS OF EXPERIMENT WITH 132 FUZES

Effect of test	Fuzes tested		Total
	At $-40°$F	At $130°$F	
Successes	64	53	117
Failures	9	6	15
Total	73	59	132

Therefore, we replace 6 by 6.5, and adjust the other frequencies to preserve the marginal totals. The adjusted data are shown in Table 4.10. If the expected frequency had been 6.4, for example, we would not have replaced 6 by 6.5; we would have accepted immediately the hypothesis of consistency, since 6 is different from 6.4 by less than ½.

TABLE 4.10. TABLE OBTAINED BY APPLYING YATES'S CORRECTION TO TABLE 4.9

Effect of test	Fuzes tested		Total
	At $-40°$F	At $130°$F	
Successes	64.5	52.5	117
Failures	8.5	6.5	15
Total	73	59	132

Using the simplified formula for a 2-by-2 table, we have

$$\chi^2 = \frac{n(a'd' - b'c')^2}{(a+b)(a+c)(c+d)(b+d)}$$

$$= \frac{132[(64.5)(6.5) - (52.5)(8.5)]^2}{(117)(73)(15)(59)} = 0.01273$$

Since the tabled value $\chi^2_{.05,1} = 3.841$ is greater than 0.01273, we cannot reject the null hypothesis that the proportion defective is the same at the two temperatures.

4.7.2. Run Test for Comparing Two Samples

The run test described in Sec. 4.1 can be used also to determine whether two random samples, say x's and y's, were drawn from populations having the same distribution. The form of the distribution need not be specified or even known.

Arrange all the readings, x's and y's together, in order of size. Count v, the number of runs in x's and y's (instead of $+$'s and $-$'s as in Sec. 4.1). Reject the null hypothesis of consistency if v is less than or equal to the value $v_{.95}$ of Appendix Table 10. (This gives a test at the 5% level of significance.)

Example. In Sec. 2.5.3b, two samples of ten were given whose respective populations were assumed to have normal distributions with the same standard deviation and with means 3.00 mm apart. Test at the 5% significance level whether the difference is 3.00 mm.

We subtract 3.00 mm from the readings of Sample 1 and list all 20 readings in order of size, as shown in Table 4.11. The number v of runs is 6. For $n_1 = n_2 = 10$, Appendix Table 10 gives $v_{.95} = 6$. Thus, we reject the null hypothesis. Note that the test says nothing about whether the distributions are normal, but merely considers whether they are the same.

TABLE 4.11. LENGTH IN MM OF
PARTS FROM TWO MACHINES

93.75	95.07[a]
93.90	95.21
94.36	95.31
94.50	95.50[a]
94.73	95.51[a]
94.82	95.60[a]
95.02	95.93[a]
95.03[a]	96.01[a]
95.06[a]	96.16
95.06[a]	96.67[a]

[a] Readings from Sample 1, with 3.00 mm subtracted.

4.8. Gross Errors

Often an experimenter wants to discard from his sample a reading that appears too large or too small, feeling that it will ruin his results. When is he justified in doing so? There is a simple answer in physical terms: He is justified whenever the reading in question did not come from the population he intended to sample but from some other population. For example, if he learns that the power supply was erratic during the period from 0840 to 0845, he is justified in throwing out readings known to have been taken during that five minutes, under the supposition that they did not come from the population with the intended power supply.

Whenever possible, such a physical criterion as this should be used when data are to be discarded. The experimenter who discards a reading just because it looks too large compared with the others may be in error, for "outside" readings are often correctly drawn from the right population and should influence the results, especially in small samples. For instance, it can be shown that in the long run one out of every 10 samples of size 3 from the same normal population will have one of the readings at least 16 times as far from the middle one as the other.

If the experimenter wants a statistical test to help satisfy himself that a reading is too extreme to have come from the right population and *if he is dealing with a normal population,* Appendix Table 16 may be used as follows: Arrange the sample in order of size, $x_1, x_2, \cdots, x_n$, starting with the smallest or the largest, depending on which value is to be considered for rejection. According to the sample size n, compute the ratio shown in Appendix Table 16. If it exceeds the critical value given under $\alpha = .05$, reject at the 5% level of significance the null hypothesis that x_1 is from the same normal population as the other sample members. In this case, omit x_1 from further calculations and use the values $x_2, x_3, \cdots, x_{n-1}, x_n$.

Example. Burning times for six firings, from a normal distribution, were recorded as follows:

0.505	0.511	0.519
0.478	0.357	0.506

Since the reading 0.357 is much smaller than the other readings, should it be discarded? Suppose the experimenter suspects that the reading was subject to an observer's error, an unusual condition in firing, or a faulty gage, but wishes to apply a statistical test to help satisfy himself that his explanation is correct.

For the six readings given above

$$r_{10} = \frac{x_2 - x_1}{x_6 - x_1} = \frac{0.478 - 0.357}{0.519 - 0.357} = 0.747$$

Since 0.747 exceeds the tabled value 0.628, listed under $\alpha = .05$, the probability is less than 5% that six readings from the same normal population would include one reading so remote from the rest.

4.9. Tolerance Intervals

To construct, on the basis of a sample n from a *normal* population, a pair of symmetric tolerance limits (Sec. 1.1.11) for $100P\%$ of the population with confidence coefficient $1 - \alpha$, obtain a coefficient K from Appendix Table 17. The desired limits are then $\bar{x} \pm Ks$, where $\bar{x}$ and s are computed from the sample. If many samples n are drawn and the limits $\bar{x} \pm Ks$ computed in each case, then $100(1 - \alpha)\%$ of the time the tolerance limits so computed will enclose at least $100P\%$ of the population from which the sample was drawn.

For more material on tolerance intervals, see Sec. 1.1.11; for more extensive tables of K, see Ref. 4.10.

Example. A sample of 28 has mean 10.02 and standard deviation $s = 0.13$. Find tolerance limits, having confidence coefficient 0.95, for 90% of the population.

For $n = 28$, $\alpha = 0.05$, and $P = 0.90$, Appendix Table 17 gives $K = 2.164$. We compute

$$\bar{x} \pm Ks = 10.02 \pm (2.164)(0.13)$$
$$= 10.02 \pm 0.28$$

The tolerance limits are 9.74 and 10.30. If tolerance limits $\bar{x} \pm 2.164s$ are constructed for many samples of size 28, in the long run 95% of such pairs will enclose at least 90% of the normal population sampled.

A one-sided upper tolerance limit U can be constructed so that with confidence coefficient $1 - \alpha$ at least $100P\%$ of the *normal* population will be less than U (Ref. 4.10). We have

$$U = \bar{x} + ks$$

where

$$k = \frac{K_{1-P} + \sqrt{K_{1-P}^2 - ab}}{a}$$

$$a = 1 - \frac{K_{a}^{2}}{2(n-1)}$$

$$b = K_{1-P}^{2} - \frac{K_{a}^{2}}{n}$$

with K_{ε} defined so that the area under the standardized normal curve to the right of K_{ε} is equal to ε as in Fig. 4.4. Similarly, a one-sided lower tolerance limit is given by $L = \bar{x} - Ks$.

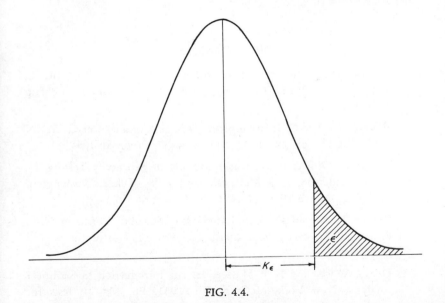

FIG. 4.4.

Example. Find a one-sided tolerance limit for the preceding example.

We read opposite $F(z) = 1 - \varepsilon$ in Appendix Table 2 the desired K_{ε}, and compute

$$K_{1-P} = K_{.10} = 1.282$$

$$K_a = K_{.05} = 1.645$$

$$n = 28$$

$$a = 1 - \frac{(1.645)^2}{2(27)} = 0.94989$$

$$b = (1.282)^2 - \frac{(1.645)^2}{28} = 1.5469$$

$$k = \frac{1.282 + \sqrt{(1.282)^2 - (0.94989)(1.5469)}}{0.94989} = 1.789$$

$$U = \bar{x} + ks = 10.02 + (1.789)(0.13) = 10.25$$

Thus, we may be 95% confident that at least 90% of the population lies below $U = 10.25$.

BIBLIOGRAPHY

4.1. Bartlett, M. S. "The Use of Transformations," BIOMETRICS BULL, Vol. 3, No. 1 (March 1947), pp. 39–51. Transformations to obtain normality.

4.2. Bliss, C. I. "The Calculation of the Dosage-Mortality Curve," ANN APPL BIOL, Vol. 22 (1935), pp. 134–67. Probit analysis.

4.3. Bureau of Ordnance. 'Staircase" Method of Sensitivity Testing, by T. W. Anderson, P. J. McCarthy, and J. W. Tukey. Washington, BuOrd, 21 March 1946. (NAVORD Report 65–46.)

4.4. Cramér, Harald. Mathematical Methods of Statistics. Princeton, N. J., Princeton University Press, 1946. Pp. 416–52. Use of the χ^2 and allied tests.

4.5. Dixon, W. J., and F. J. Massey, Jr. An Introduction to Statistical Analysis. New York, McGraw-Hill, 1951. Pp. 184–91, tests for goodness of fit and independence; pp. 254–57, the run test and confidence bands for cumulative distribution curves; and pp. 279–87, up-and-down method of sensitivity testing.

4.6. Finney, D. J. Probit Analysis, 2nd ed. Cambridge, England, Cambridge University Press, 1952.

4.7. Fisher, R. A. Statistical Methods for Research Workers, 10th ed. Edinburgh, Scotland, Oliver & Boyd, Ltd., 1946. Sec. 21.02.

4.8. Latscha, R. "Tests of Significance in a 2-by-2 Contingency Table: Extension of Finney's Table," BIOMETRIKA, Vol. 40 (1953), pp. 74–86.

4.9. Pearson, E. S., and H. O. Hartley. Biometrika Tables for Statisticians, Vol. 1. Cambridge, England, Cambridge University Press, 1954. Table 38, pp. 188–93.

4.10. Statistical Research Group, Columbia University. Selected Techniques of Statistical Analysis, ed. by Churchill Eisenhart, M. W. Hastay, and W. A. Wallis. New York, McGraw-Hill, 1947. Pp. 46–47, one-sided tolerance limits; pp. 95–110, symmetric tolerance limits; and pp. 249–53, contingency tables.

4.11. U. S. Naval Ordnance Laboratory. Statistical Methods Appropriate for Evaluation of Fuze Explosive-Train Safety and Reliability, by H. P. Culling. White Oak, Md., NOL, 13 October 1953. (NAVORD Report 2101.)

Chapter 5

PLANNING OF EXPERIMENTS AND
ANALYSIS OF VARIANCE

5.0. Introduction

An experiment is generally conducted (1) to determine whether some conjectured effect exists, and (2) if the effect does exist, to determine its size. If the experimenter is interested only in the effect of one **factor** (or variable) on one other variable, the design of his experiment may·be simple. But if he is concerned with more than one factor, as is frequently the case, his experiment or experiments will be more complex and may be conducted in various ways. The classical method is to restrict attention to one variable at a time, the other factors being held fixed; however, this method also restricts conclusions. When several factors are of interest, more reliable and more general conclusions can be reached by introducing further **levels** of these factors in the initial design.

Allowing more than one factor to vary in a single integrated design also permits determination of the **interaction** of factors; for instance, we may learn that increasing a constituent of propellant powder increases the burning rate more at one powder temperature than at another.

Aside from the factors that we want to investigate, there are background conditions which may affect the results of the experiment. Some of them may be taken into account explicitly in the design. The influence of the others should be minimized by scheduling the experiments with all the desired combinations of levels and factors in a random order that has been determined, for instance, by consulting a table of random numbers.

There may be variations in background conditions that are unknown to the experimenter. To be able to conclude validly that a postulated effect exists, he should plan the experiment so that a control item will be subjected to the same background conditions (except for random fluctuations) as each experimental item. For example, it cannot be concluded that a new type of rocket has a smaller dispersion than a previous type unless the two are fired under the same conditions; for instance, one rocket of the previous type might be fired as a control with each new rocket.

Measurements are not precisely reproducible; therefore, to an extent depending on the size of the experimental error, the experiment must be repeated in order to draw valid conclusions. Each complete repetition of an experiment is called a **replication**. The most efficient procedure would be to (1) estimate the experimental error before the experiment on the basis of previous experience, (2) determine the number of replications accordingly, and (3) perform all the various parts of the total experiment in a random order.

The reliability of experimental conclusions can also be increased by refining the experimental technique (reducing the standard deviation of the experimental error). Since a few replications of a refined technique can achieve the same reliability as many replications of a coarse technique, the choice of method in a particular investigation may be made on the basis of cost.

These general principles are exemplified in later sections. However, there are many designs more complex than those given in this Manual; for these, the experimenter should consult a statistician or study the references at the end of this chapter.

The data obtained from an experiment involving several levels of one or more factors are analyzed by the technique of **analysis of variance**. This technique enables us to break down the variance of the measured variable into the portions caused by the several factors, varied singly or in combination, and a portion caused by experimental error. More precisely, analysis of variance consists of (1) a partitioning of the *total* sum of squares of deviations from the mean into two or more *component* sums of squares, each of which is associated with a particular factor or with experimental error, and (2) a parallel partitioning of the total number of degrees of freedom.

Let us consider an experiment on the extrusion of propellant grains in which we desire to determine the effects of extrusion rate and die temperature on grain port area. Ambient temperature and humidity during extrusion are background conditions that may affect the results. Values of port area might be observed at several levels of each factor, at extrusion rates of 15, 20, and 25 in/min and at die temperatures of 120 and 140°F. Each combination of one level of extrusion rate with one level of die temperature is called a **treatment**; for instance, an extrusion rate of 20 in/min with a die temperature of 140°F. (Some authors call this a treatment combination rather than a treatment. The levels of each factor then become the treatments.) If the background conditions were controlled at each of several levels, each treatment would also include one level of ambient temperature and one level of humidity. Alternatively, ambient temperature and humidity might be (1) held constant (thus restricting the range of the conclusions); (2) allowed to vary as usual, but with their effects randomized by randomizing the order of treatments; or (3) recorded during the experiment and later treated by analysis of covariance. (For information on the analysis of covariance, see Ref. 5.5, 5.14, and 5.15.)

Analysis of variance, analysis of covariance, and regression analysis are specializations of the more general theory of testing linear hypotheses. (For an explanation of regression analysis, see Chapter 6 of this Manual; for the theory of testing linear hypotheses, see Ref. 5.10, 5.11, and 5.12.)

5.1. Common Designs for Experiments

The following considerations apply generally in designing experiments or comparing designs. Some specific designs are treated in Sec. 5.1.1–5.1.4.

The precision of an experiment is measured in terms of σ_0, the standard deviation (assumed common) of the populations of observations for each treatment—in other words, the experimental error. (See Sec. 5.3.) If the design is given m replications, so that each treatment is applied m times, the standard deviation of the estimate of the difference between two treatment effects is proportional to $\sigma_0/\sqrt{m}$. Thus, as the number of replications increases, the error decreases. This formula applies only if the replications are independent and the experimental material remains homogeneous as the experiment increases in size. Elaborations of such error considerations make it possible to determine how large an experiment should be for detection of effects of a prescribed size, somewhat in the manner that Appendix Charts V–X can be used in simpler experiments. (See Ref. 5.4 and 5.10.)

Designs may be compared on the basis of the number of degrees of freedom for estimating the standard error σ_0. A decrease in the number of degrees of freedom will decrease the sensitivity of the experiment. Suppose that an observed variable, compressive strength of grain, is to be tested with four levels each of three factors: temperature, extrusion rate, and humidity. Let us consider the following methods:

1. We may use a full-scale $4 \times 4 \times 4$ factorial design (Sec. 5.1.1) with 64 treatments. This will yield 27 degrees of freedom for estimating the error variance.

2. We may decide to include only two of the three factors if the results obtained from these are to be applied to only one level of the third factor, or if the four levels of the third factor are expected to have about the same effect on the compressive strength of the grain. (If this assumption is not justified, the standard error σ_0 will be increased by the variation among the levels in case the four levels of the third factor are included by randomization. If only one level of the third factor is included, the results will apply only to that level.) Using a 4×4 factorial design with 16 treatments will yield nine degrees of freedom for estimating the error variance if the two factors are independent; i.e., if their interaction can be neglected.

3. We may replicate the 4×4 design of paragraph 2 four times to give 64 applications of the 16 treatments (four each). This procedure will yield 48 degrees of freedom for estimating the error variance. Then the standard error for the difference between two treatment effects will be the same as in paragraph 1 and one-half that in paragraph 2.

4. We may use a 4×4 Latin-square arrangement (Sec. 5.1.4) of two factors, with the four levels of the third factor superimposed according to a randomly selected 4×4 Latin square. This design assumes that all three factors are independent and gives information on all three with a minimum number of treatments—16 in the example. However, there are only six degrees of freedom for estimating the error variance; as a result, substantial real effects of the factors may go undetected because of the large random fluctuation which must be allowed for error.

Which of the above designs is best cannot be decided categorically without further knowledge of the conditions to be met. Under various conceivable circumstances any one of these designs, or some other design, might be chosen. Whatever the bases for decision, it is essential that a systematic method be used for selecting and carrying out the design. The following list provides a useful general routine for designing and running an experiment.

1. *State the objectives of the experiment.*
   a. Use information from previous experiments.
   b. Choose experimental conditions to represent the conditions under which the results will be applied. This may be done by either systematic choice or randomization.
   c. State the precise hypotheses that are to be tested.
2. *Draw up a preliminary design.*
   a. Take into account the experimental error, and the number of degrees of freedom for estimating it, provided by each proposed design.
   b. Consider the cost of experimentation versus the cost of wrong decisions.
3. *Review the design with all collaborators.*
   a. Reach an understanding as to what decisions hinge on each outcome. Keep notes.
   b. Encourage collaborators to anticipate all factors that might affect the results.
   c. Discuss the experimental techniques in sufficient detail to discover any procedures that might lead to bias.
4. *Draw up the final design.*
   a. Present the design in clear terms to assure that its provisions can be followed without confusion.
   b. Include the methods of analysis as part of the design, ascertaining that conditions necessary for the validity of these methods will be met.
5. *Carry out the experiment.*
   a. During the course of the experiment, maintain communication among all collaborators, so that questions arising from unforeseen experimental conditions or results may be answered in keeping with the design agreed upon.
6. *Analyze the data.*
   a. Follow the methods outlined in the final design (Step 4).
7. *Write a report.*
   a. Present the data and results in clear tables and graphs.

b. Compare the results with the stated objectives of the experiment (Step 1).

c. If the results suggest further experimentation, outline the course that such experimentation should take.

5.1.1. Complete Factorial Design for Treatments

The complete factorial design for arranging the treatments to be tested calls for the use of every combination of the different levels of the factors. Applying the complete factorial design to the example in Sec. 5.0, the number of treatments would be $3 \times 2 = 6$, as shown in Table 5.1.

TABLE 5.1. TREATMENTS FOR COMPLETE FACTORIAL
DESIGN OF PROPELLANT-GRAIN EXPERIMENT

Treatment	Extrusion rate, in/min	Die temperature, °F
1	15	120
2	15	140
3	20	120
4	20	140
5	25	120
6	25	140

Note that in this design both factors, extrusion rate and die temperature, are investigated simultaneously, whereas in the classical method all variables but one remain fixed. The advantages of complete multiple factorial design are (1) the increased generality of the conclusions for each factor, which may apply to several levels of the other factors rather than to one fixed level; and (2) the possibility of checking for *interactions* (i.e., effects produced by two or more factors jointly that go beyond the total of their individual effects). Examples of interaction are common in chemistry: the properties of a mixture may differ widely from the properties of the components.

To compare only two treatments—an extrusion rate of 15 in/min and a die temperature of 120°F with an extrusion rate of 20 in/min and a die temperature of 120°F—and to estimate the experimental error by the variability within each treatment, we could use the *t* test (Sec. 2.5.3b).

Also, all six treatments mentioned above could be compared pairwise in $(6 \times 5)/2 = 15$ t tests, but these tests would not be independent. More information can be extracted from the data if we study all the treatments together.

5.1.2. Complete Randomization for Background Conditions

Suppose we wish to test the effects of the use of Friden, Marchant, and Monroe desk computers (three treatments) on the time it takes to compute sums of squares of 12 four-digit numbers, and to obtain conclusions unaffected by the variability among operators.

Several background conditions, though not of prime importance in the experiment at hand, are likely to influence the results. If only one operator is used, he may have a natural preference for one of the machines, so that the experimental results will apply only to him. If several operators are used, one may be so much more skilled than the others that his computer will appear best even if it is not. The order of the trials may make a difference. Each operator may become more relaxed during the test and his speed may consistently increase as the test progresses, or he may tire and his speed may consistently decrease. If he uses the same twelve numbers in each trial, his proficiency will almost certainly increase.

Some of these conditions may be varied systematically. For instance, we can arrange the experiment so that each operator will use each machine for the same number of trials. Other conditions, such as the order of trials, can be randomized by flipping coins or by drawing numbered chips from a bowl. This procedure may still favor some particular machine (by placing it last, say) unless the entire experiment is repeated several times to yield several replications. Familiarity with the same numbers can be avoided by drawing anew 12 four-digit numbers from a table of random numbers for each operator-machine combination.

Successful experiments depend heavily on ingenuity in spotting possible sources of bias and eliminating them by a well-organized plan. *Before the experiment*, we should decide upon and record the procedure that will be followed for every contingency that may arise during the experiment, and the conclusion that will be drawn from every possible outcome. In the computing experiment, for instance, we should decide on the procedure in case one of the sums is computed incorrectly, or if a claim is presented later that one of the machines was out of adjustment.

Complete randomization is recommended for small experiments on material known to be homogeneous and, because of ease in analysis compared with other designs, for any experiment likely to contain failures that must be omitted from the analysis.

5.1.3. Randomized-Block Design for Background Conditions

In larger experiments the background conditions may lack homogeneity, so that complete randomization of the whole background at once would introduce an unnecessary amount of error. In this case, we may decide to randomize one *block* of the background at a time, keeping each block intact so that its effect will appear explicitly in the analysis.

In the test of desk computers described above, considerable variation might occur if an operator were trained on a particular machine, or had used it the most, or preferred it. Restricting the test to operators trained on Fridens would limit our conclusions to such operators. Randomizing the selection of operators according to their training and analyzing the total data would not ascertain whether the training of the operator had affected the results; if there was such an effect it would appear as experimental error. We might prefer to consider the operators trained on the three different machines in three separate blocks, preferably of equal size. Within each block we would still randomize the order of use of the computers. At the end of the experiment we could assess the effect of the type of training on computing speed.

5.1.4. Latin-Square Design for Background Conditions

To test k treatments under k choices for one background condition and k choices for a second, independent background condition, we can use a Latin-square design; i.e., a square array of letters such that each letter appears exactly once in each row and exactly once in each column. (For further material on Latin squares, see Ref. 5.9; for examples of Latin squares of various sizes, see Ref. 5.7.)

Suppose we wish to test the effect of office noise level on the speed of computing, taking into account as background conditions the desk computer used and the operator making the trial. We might choose four supposedly like computers at random from the entire production of a given model and four operators at random from a large office staff. The trials might then be arranged in a Latin square as in Table 5.2, operator 1 computing on machine 3 at noise level D, for instance. The order of the trials could be randomized to take care of the learning factor. Only $4 \times 4 = 16$ tests are required. If more than 16 tests are desired, further randomly selected Latin squares of the same size may be used.

TABLE 5.2. A 4×4 LATIN SQUARE FOR NOISE-LEVEL EXPERIMENT

The letters A, B, C, and D represent levels of noise, the volume increasing from A to D.

Background condition II (operator)	Background condition I (computer)			
	1	2	3	4
1	B	C	D	A
2	D	B	A	C
3	C	A	B	D
4	A	D	C	B

Advantages of the Latin-square design are economy of samples and ready analysis. The main disadvantage is inflexibility, since the same number of choices is required for each variable and the number of degrees of freedom is thereby determined also.

The restriction that the two background conditions and the factor to be tested be independent (no interactions) is sometimes a hard one to meet. For instance, in the noise-level experiment it must be assumed that the level of noise has the same effect on the speed of computing, regardless of the operator. If one operator is particularly bothered by noise or another operator is unusually oblivious to noise while he computes, the conclusions of the experiment will be made unreliable.

5.1.5. Other Designs

Generalizations of these designs, such as the Greco-Latin square (k levels of each of three background conditions), incomplete block designs, and designs using "confounding" including those using fractional replication, are available to meet specific needs (Ref. 5.4, 5.10, and 5.15). For the analysis of these and other basic designs, see Ref. 5.10.

Most designs have symmetries that make their analysis simple. However, designs that are unsymmetrical (because data are missing, for example) may be analyzed according to the general theory of testing linear hypotheses (Ref. 5.11). This theory has become somewhat easier to apply since the advent of high-speed electronic computing machines.

5.2. Types of Analysis of Variance

There are two mathematical models to which the analysis of variance applies—a model for Type I problems (Sec. 5.2.1) and a model for Type II problems (Sec. 5.2.2). This Manual does not treat the more complicated case of "mixed" models; material on this subject may be found in Ref. 5.1 and 5.15.

5.2.1. Type I Problems

When particular levels of a factor are selected purposely for study because they are the only levels of interest or are considered the most promising, the analysis of variance becomes a comparison of the mean effects of those particular levels. Statistical tests (F tests) are made to see whether the observed differences in mean effects are real or random. If the differences are real, the population constants or parameters (main effects and interactions) may be estimated easily as averages. In these Type I problems, the analysis of variance refers to the finite number of treatments actually chosen, and the variation is called **systematic**. The problems treated in most elementary texts are of this type (Ref. 5.5). The experiments on desk computers described in Sec. 5.1.2 and 5.1.3 involve the systematic variation of machines from Friden to Marchant to Monroe.

5.2.2. Type II Problems

When levels of each factor are drawn at random, the analysis of variance is concerned not so much with the particular levels appearing in the experiment as with the larger finite or infinite population of levels that had an equal chance of being drawn. The analysis of variance provides statistical tests (F tests) to see whether the random variation of levels of each factor actually contributes variation to the variable under study. These tests for Type II problems are of the same form as those used for Type I problems; however, the objective in Type II problems is the estimation of the component of variance contributed by each factor to the total variance of the variable under study, rather than the estimation of the main effects and interactions at particular levels of each factor, as in Type I problems (Ref. 5.6).

In the noise-level experiment of Sec. 5.1.4 we drew four computers at random from the total production of one model and four operators at random from a large office staff. Here, then, conclusions as to the effect of noise level can be drawn for *all* computers of that model and for *all* operators on the staff. The experiment will also show the variability of computers and the variability of operators in their effects on speed of computing.

5.3. Assumptions of Analysis of Variance

(1) Observations are random with respect to any conditions not systematically varied or removed from the analysis by covariance methods.

(2) Means and variances are additive. In the Type I model we look upon each observation (x_{ijt}, for example, which receives factor 1 at level i and factor 2 at level j) as composed of the separate parts

$$x_{ijt} = \mu + \alpha_i + \beta_j + e_{ijt}$$

where

$\mu =$ over-all mean

$\alpha_i =$ mean effect of factor 1 at level i

$\beta_j =$ mean effect of factor 2 at level j

$e_{ijt} =$ random deviation from the mean position of the tth item receiving the treatment ij. The e_{ijt} are assumed to have population means of zero.

In the Type I model $\sum_i \alpha_i = \sum_j \beta_j = 0$.

In the Type II model we again take the observation x_{ijt} as a sum of the terms given above, but now every term except μ is random, so that the variance of a random observation x_{ijt} is the sum of a variance component caused by factor 1, a variance component caused by factor 2, and an error variance not caused by either factor. In many problems the variation caused by several factors is greater than the sum of their independent effects, so that other joint effects (interactions) must be added, as in Sec. 5.5.2.

(3) Experimental errors e_{ijt} are independent.

(4) Variances of the experimental errors e_{ijt} for all pairs i, j are equal, with the common value σ_0^2.

(5) Distribution of experimental errors is normal.

To estimate mean effects or components of variance, we need only the first four assumptions. To make significance tests of the means or components, or to determine how precise (or good) our estimates are, we must satisfy all five assumptions. If certain of the assumptions are not met, the analysis of variance provides only an approximate procedure; however, this approximation is usually the best available. (See Ref. 5.3 for a discussion of the effects of deviations from the assumptions, and Ref. 5.2 for methods of inducing normality.)

5.4. One-Factor Analysis

Table 5.3 shows in general form the tabulation that would result from an experiment to compare r different treatments, levels of a single factor. Each treatment has been used (replicated) m times. There is some random variation among the readings for any one treatment, but it is the treatment sample means that are compared. The difference among the means is not considered significant unless it is large compared with the random within-treatments variation.

Throughout Sec. 5.4.1 and 5.4.2 the reader may find it helpful to follow the examples given in Sec. 5.4.2 and to compare Table 5.5 with Table 5.3 and Tables 5.6 and 5.7 with Table 5.4.

5.4.1. Assumptions

For general assumptions of the analysis of variance, see Sec. 5.3. The assumptions for one-factor analysis follow.

(1) The m replications $x_{i1}, x_{i2}, \ldots, x_{im}$ of the experiment for any one treatment i represent a sample drawn at random from a normal population (see Table 5.3). Each observation x_{it} can be written as

$$x_{it} = \mu + \alpha_i + e_{it} \qquad (i = 1, 2, \cdots, r; t = 1, 2, \cdots, m)$$

where

μ = over-all mean

α_i = mean effect of the ith treatment (or row)

e_{it} = random deviation from the mean position of the tth item receiving treatment i. The e_{it} have independent normal distributions with mean zero and common variance σ_0^2.

In the Type I model, $\sum_i \alpha_i = 0$.

In the Type II model we assume that the variance of the population of row effects is σ_a^2.

(2) For a Type I problem, the r treatments are chosen systematically by the experimenter as important levels of the factor. For a Type II problem, on the contrary, we assume that the r levels of the factor are chosen at random from a normal population of levels of the factor.

TABLE 5.3. ONE-WAY CLASSIFICATION

Treatment	Replications					
	1	2	$\cdots$	t	$\cdots$	m
1	x_{11}	x_{12}	$\cdots$	x_{1t}	$\cdots$	x_{1m}
2	x_{21}	x_{22}	$\cdots$	x_{2t}	$\cdots$	x_{2m}
$\vdots$	$\vdots$	$\vdots$	$\vdots$	$\vdots$	$\vdots$	$\vdots$
i	x_{i1}	x_{i2}	$\cdots$	x_{it}	$\cdots$	x_{im}
$\vdots$	$\vdots$	$\vdots$	$\vdots$	$\vdots$	$\vdots$	$\vdots$
r	x_{r1}	x_{r2}	$\cdots$	x_{rt}	$\cdots$	x_{rm}

r = number of tested levels of the factor (number of rows)

m = number of replications of each treatment (number of measurements per treatment)

$n = rm$ = total number of observations

x_{it} = replication t of treatment i (trial t in row i)

$x_{i.} = \dfrac{1}{m} \sum_{t=1}^{m} x_{it}$ = ith treatment mean (mean of the ith row)

$\bar{x} = \dfrac{1}{r} \sum_{i=1}^{r} x_{i.} = \dfrac{1}{n} \sum_{i,t} x_{it}$ = grand mean

5.4.2. Analysis

The sum of the squared deviations of all $n = rm$ observations from the grand mean $\bar{x}$ can be written as m times the sum of squared deviations of the r treatment means from the grand mean plus the sum of squared deviations of individual observations from their respective treatment means

$$\sum_{i=1}^{r} \sum_{t=1}^{m} (x_{it} - \bar{x})^2 = m \sum_{i=1}^{r} (x_{i.} - \bar{x})^2 + \sum_{i=1}^{r} \sum_{t=1}^{m} (x_{it} - x_{i.})^2$$

In the case of a Type I problem we make the null hypothesis that the means of the normal populations of observations x_{it} for the r treatments are all equal; that is, that the α_i are all zero. If the null hypothesis is true, we have three unbiased estimates of the common variance σ_0^2 of the r populations, obtained by dividing the sums of squares in the equation above by their respective degrees of freedom. The estimates are

$$s_1^2 = \frac{1}{n-1} \sum_i \sum_t (x_{it} - x)^2$$

$$s_2^2 = \frac{m}{r-1} \sum_i (x_{i.} - x)^2$$

$$s_3^2 = \frac{1}{n-r} \sum_i \sum_t (x_{it} - x_{i.})^2$$

(but the algebraically equivalent computing forms in Table 5.4 are preferable for numerical work).

TABLE 5.4. ANALYSIS OF VARIANCE FOR ONE-WAY CLASSIFICATION

Source of variation	Sum of squares	Degrees of freedom	Mean square	Expected mean square
Among treatments	$(2) = \dfrac{r \sum_i \left(\sum_t x_{it} \right)^2 - \left(\sum_{i,t} x_{it} \right)^2}{n}$	$r-1$	$s_2^2 = \dfrac{(2)}{r-1}$	$m\sigma_a^2 + \sigma_0^2$
Within treatments	$(3) = (1) - (2)$	$n-r$	$s_3^2 = \dfrac{(3)}{n-r}$	σ_0^2
Total	$(1) = \dfrac{n \sum_{i,t} x_{it}^2 - \left(\sum_{i,t} x_{it} \right)^2}{n}$	$n-1$		

If the null hypothesis of equal means is true, the last two quantities should be about equal (except for sampling error) with a ratio near one, for they represent independent estimates of σ_0^2. We divide the second by the third, and reject the null hypothesis at the significance level α if the quotient F exceeds the critical F value from Appendix Table 5 for $f_1 = r - 1$ and $f_2 = n - r$ degrees of freedom. A value for F greater than or equal to the tabled value $F_\alpha(r - 1, n - r)$ would occur in random sampling with probability only α if the null hypothesis were true. For the F tests in this chapter, we are interested in only one tail of the distribution, since the variability we want to detect can only tend to increase the among-treatments variance, never to decrease it.

If the null hypothesis is false, only s_3^2 of the three mean squares is an unbiased estimate of σ_0^2. In this case, the x_i. are the best estimates of the different population means $\mu + \alpha_i$.

If the quotient F does not exceed $F_\alpha(r - 1, n - r)$, the data are consistent with the null hypothesis that the means are all equal. In this case the sample grand mean $\bar{x}$ is taken as the best estimate of the common population mean μ. The calculations of the three estimates of σ_0^2 can be made conveniently as shown in Table 5.4. The last column of the table is appropriate only in Type II problems.

As is indicated by comparison of the last two columns, s_3^2 estimates the variance σ_0^2 caused by variability within the individual treatments; and if the treatments represent a random selection from the population of levels taken on by the factor (Type II problem), the quantity

$$s_a^2 = \frac{s_2^2 - s_3^2}{m}$$

estimates the variance σ_a^2 caused by variability among the treatments (variability among rows). Notice that the estimate s_2^2 obtained from considering the dispersion of the r treatment means includes some fluctuation caused by within-treatment variability, which the operation expressed by the equation above removes, in accordance with the last column of the table.

The expressions in the second column of the table (the sums of squares) have been abbreviated to (1), (2), and (3) for convenience.

The estimate $s_1^2 = (1)/(n - 1)$ of the variance could be formed also, but is dependent on (2) and (3), which answer directly the usual questions of interest.

Unequal sample sizes m_i for the different treatments are allowable, but not recommended. The analysis is the same except for the formula

$$(2) = \sum_i \frac{\left(\sum_t x_{it}\right)^2}{m_i} - \frac{\left(\sum_{i,t} x_{it}\right)^2}{n}$$

where n is the total number of observations, $\sum_{i=1}^{r} m_i$.

The following example could be either Type I or Type II, according to the circumstances. The analysis is in great part the same for both, and will be carried out for both types.

Example (Type I Problem). The data of Table 5.5 (a specific case of the tabular form shown in Table 5.3) represent coded readings on widths of propellant carpet rolls from 10 production lots, with 12 samples selected randomly from each lot. The *factor* to be considered for its effect on width of roll is the lot. The $r = 10$ levels of the factor are the treatments, each of which received $m = 12$ replications. The lots show different mean sample widths. Is this difference large enough compared with the variability within lots to indicate a real difference among lots?

TABLE 5.5. One-Way Classification of Propellant
Carpet-Roll Data

Lot	Samples											
	1	2	3	4	5	6	7	8	9	10	11	12
1	40	30	38	23	35	38	52	59	28	37	44	54
2	52	44	47	3	31	2	2	50	59	32	2	60
3	59	5	3	5	44	8	46	0	19	19	15	29
4	44	12	23	21	32	6	40	3	52	32	29	44
5	50	35	50	25	4	50	40	3	32	28	22	54
6	20	55	31	30	40	40	42	46	30	52	15	46
7	19	58	40	8	54	38	56	45	50	24	21	0
8	57	7	46	33	4	19	7	38	47	52	24	42
9	10	42	24	9	54	35	18	16	50	48	38	15
10	49	7	43	33	39	2	3	10	25	6	3	38

NOTE: $r = 10$ treatments; $m = 12$ replications; $n = 120$ observations; $\bar{x} = 30.8$.

TABLE 5.6. CALCULATIONS ON CARPET-ROLL DATA

In using a machine computing technique, only the total, not the separate entries, for the column of squared sums would be tabulated.

Lot	$\sum_t x_{it}$	$\left(\sum_t x_{it}\right)^2$	$\sum_t x_{it}^2$
1	478	228,484	20,332
2	384	147,456	18,436
3	252	63,504	9,444
4	338	114,244	12,224
5	393	154,449	16,183
6	447	199,809	18,311
7	413	170,569	18,427
8	376	141,376	15,626
9	359	128,881	13,715
10	258	66,564	9,136
Total	3,698	1,415,336	151,834

$$(1) = \frac{n \sum_{i,t} x_{it}^2 - \left(\sum_{i,t} x_{it}\right)^2}{n} = \frac{120(151,834) - (3,698)^2}{120} = 37,873.967$$

$$(2) = \frac{r \sum_i \left(\sum_t x_{it}\right)^2 - \left(\sum_{i,t} x_{it}\right)^2}{n} = \frac{10(1,415,336) - (3,698)^2}{120} = 3,984.633$$

$$(3) = (1) - (2) = \frac{4,544,876 - 478,156}{120} = 33,889.333$$

For a Type I problem, the 10 lots of carpet rolls in Table 5.5 are assumed to be lots of particular interest specified by the experimenter, say the first 10 lots produced. We assume (Sec. 5.4.1) that the widths for each lot are distributed normally with variance σ_0^2.

From the data of Table 5.5 we form the sums of Table 5.6. (In using a machine computing technique, only the total, not the separate entries, for the column of squared sums would be tabulated. The separate squares are shown to accustom the reader to the summation notation.) Table 5.7 is our analysis-of-variance table.

To test the null hypothesis that the 10 treatment means are equal, we compute the quotient F of the among-treatments mean square by the within-treatments mean square. To make the test at the 5% significance level, we compare

$$F = \frac{442.737}{308.085} = 1.437$$

with $F_a(r - 1, n - r) = F_{.05}(9, 110)$. From Appendix Table 5 we find that $F_{.05}(9, 110)$ is at least 1.96. Since 1.437 is less than $F_{.05}(9, 110)$, we cannot reject the null hypothesis that the lot means are equal; in other words, the data are consistent with the null hypothesis.

Analysis of variance in the example above is a way of determining whether the production process is in statistical control; a control chart on means answers the same question more simply and rapidly, though more roughly (see Chapter 7).

Example (Type II Problem). In the preceding example, the 10 lots were specified by the experimenter. They could have been any 10 specific lots of interest to him.

Suppose that 10 lots are not specified, but selected at random from a population of lots whose mean roll widths are distributed with variance σ_a^2, and that the widths within each lot have a normal distribution with variance σ_0^2. Now we have a Type II problem and we may extend the analysis to include components of variance. Our object is to estimate the between-lots variance σ_a^2 (by s_a^2), and the within-lots variance σ_0^2 (by s_3^2).

Using Tables 5.4 and 5.7, we find the estimates

$$s_3^2 = 308.085 \qquad \text{and} \qquad ms_a^2 + s_3^2 = 442.737$$

Then

$$s_a^2 = \frac{442.737 - 308.085}{12} = 11.221$$

The estimated standard deviation caused by variability among all lots of the population is $\sqrt{11.221} = 3.350$. The estimated standard deviation

TABLE 5.7. ANALYSIS OF VARIANCE FOR CARPET-ROLL DATA

Source of variation	Sum of squares	Degrees of freedom	Mean square
Among treatments	3,984.633	9	442.737
Within treatments	33,889.333	110	308.085
Total	37,873.967	119	

caused by variability within the individual lots is $\sqrt{308.085} = 17.552$. The estimated total variance in carpet-roll widths in the population of lots is

$$s_a^2 + s_3^2 = 11.221 + 308.085 = 319.306$$

5.5 Two-Factor Analysis

5.5.1. Two-Factor Analysis Without Replication

To investigate r levels of factor 1 and c levels of factor 2, we may take the $n = rc$ observations shown in Table 5.8. We have the underlying assumption that each observation may be written

$$x_{ij} = \mu + \alpha_i + \beta_j + e_{ij} \qquad (i = 1, 2, \cdots, r; j = 1, 2, \cdots, c)$$

with

$$\sum_i \alpha_i = \sum_j \beta_j = 0$$

in the Type I model; i.e., it is the sum of an over-all mean, a row effect, a column effect, and a random error. It is also assumed that the errors e_{ij} are independently and normally distributed, with mean zero and common variance σ_0^2.

Only one observation is taken for each treatment; i.e., the experiment is not replicated. There are $n = rc$ treatments—each level of factor 1 is used once with each level of factor 2. The typical analysis-of-variance table is shown in Table 5.9. The last column, indicating the components of variance σ_a^2, σ_β^2, and σ_0^2, applies only to Type II problems. We notice that, in addition to the inclusion of the among-columns values, Table 5.9 differs from Table 5.4 in that it gives "residual" values instead of within-treatments values. Since here each of the rc treatments receives only one trial, we do not have a direct estimate of the variance of a whole population of observations of that treatment. However, under the assumptions, this variance is σ_0^2 and can be estimated by the residual mean square. The row or column mean square provides an estimate of σ_0^2 only if the corresponding null hypothesis that every $\alpha_i = 0$, or that every $\beta_j = 0$, is true. Hence, to test the null hypotheses, we compare the row and column mean squares with the residual mean square.

TABLE 5.8. Two-Way Classification (Without Replication)

Only the total, not the separate squares, of the next-to-last row and column need be tabulated.

Factor 1	Factor 2						$\sum_j x_{ij}$	$(\sum_j x_{ij})^2$	$\sum_j x_{ij}^2$
	Level 1	Level 2	...	Level j	...	Level c			
Level 1	x_{11}	x_{12}	...	x_{1j}	...	x_{1c}	$\sum_j x_{1j}$	$(\sum_j x_{1j})^2$	$\sum_j x_{1j}^2$
Level 2	x_{21}	x_{22}	...	x_{2j}	...	x_{2c}	$\sum_j x_{2j}$	$(\sum_j x_{2j})^2$	$\sum_j x_{2j}^2$
...	...	...	...	...	...	...	...	...	...
Level i	x_{i1}	x_{i2}	...	x_{ij}	...	x_{ic}	$\sum_j x_{ij}$	$(\sum_j x_{ij})^2$	$\sum_j x_{ij}^2$
...	...	...	...	...	...	...	...	...	...
Level r	x_{r1}	x_{r2}	...	x_{rj}	...	x_{rc}	$\sum_j x_{rj}$	$(\sum_j x_{rj})^2$	$\sum_j x_{rj}^2$
$\sum_i x_{ij}$	$\sum_i x_{i1}$	$\sum_i x_{i2}$	...	$\sum_i x_{ij}$	...	$\sum_i x_{ic}$	$\sum_i \sum_j x_{ij}$		
$(\sum_i x_{ij})^2$	$(\sum_i x_{i1})^2$	$(\sum_i x_{i2})^2$	...	$(\sum_i x_{ij})^2$	...	$(\sum_i x_{ic})^2$	$\sum_i (\sum_j x_{ij})^2$		
$\sum_i x_{ij}^2$	$\sum_i x_{i1}^2$	$\sum_i x_{i2}^2$	...	$\sum_i x_{ij}^2$	...	$\sum_i x_{ic}^2$			$\sum_{ij} x_{ij}^2$

TABLE 5.9. ANALYSIS OF VARIANCE FOR TWO-WAY CLASSIFICATION (WITHOUT REPLICATION)

Source of variation	Sum of squares	Degrees of freedom	Mean Square	Expected mean square
Among factor 1 levels (among rows)	$(2) = \dfrac{r\sum_i\left(\sum_j x_{ij}\right)^2 - \left(\sum_{i,j} x_{ij}\right)^2}{n}$	$r - 1$	$s_2^2 = \dfrac{(2)}{r-1}$	$c\sigma_a^2 + \sigma_0^2$
Among factor 2 levels (among columns)	$(3) = \dfrac{c\sum_j\left(\sum_i x_{ij}\right)^2 - \left(\sum_{i,j} x_{ij}\right)^2}{n}$	$c - 1$	$s_3^2 = \dfrac{(3)}{c-1}$	$r\sigma_\beta^2 + \sigma_0^2$
Residual	$(4) = (1) - (2) - (3)$	$(r-1)(c-1)$	$s_4^2 = \dfrac{(4)}{(r-1)(c-1)}$	σ_0^2
Total	$(1) = \dfrac{n\sum_{i,j} x_{ij}^2 - \left(\sum_{i,j} x_{ij}\right)^2}{n}$	$n - 1$		

To test at the significance level 0.05, for example, the null hypothesis that all row means are equal (that factor 1 has no effect), we compare

$$F = \frac{s_2^2}{s_1^2}$$

with $F_{.05}[r - 1, (r - 1)(c - 1)]$ from Appendix Table 5, and reject the null hypothesis if F exceeds the tabled value. To test the null hypothesis that the column means are equal (that factor 2 has no effect), we compare

$$F = \frac{s_3^2}{s_1^2}$$

with $F_{.05}[c - 1, (r - 1)(c - 1)]$. The significance test for row effects is "orthogonal to" the test for column effects; i.e., the presence or absence of column effects does not influence the test for row effects (and vice versa), but the two test ratios have the same denominator and as a result are not statistically independent. (See Ref. 5.12, p. 321.)

In the case of a Type II problem (Sec. 5.2.2), the expected-mean-square column set equal, term by term, to the mean-square column can be used to obtain

$s_1^2 =$ estimate of the variance σ_0^2 not caused by either factor by itself (Ref. 5.12, pp. 342–44)

$s_a^2 =$ estimate of the variance σ_a^2 caused by factor 1

$s_\beta^2 =$ estimate of the variance σ_β^2 caused by factor 2

Example (Type I Problem). Table 5.10 shows the results of an experiment to determine the effects of two factors (volume of sample and operator) on the variable (percentage of nitroglycerin) in a chemical determination. The experiment is designed to test whether the chemical determination of the percentage of nitroglycerin in a solution depends on the volume of the sample drawn for analysis.

The data have been coded. Three levels of operator were chosen (three specific operators), and each operator made determinations on five samples of different volumes. In Table 5.10 only the total, not the separate squares, in the next-to-last row and column need be tabulated. Table 5.11 gives the analysis of variance.

TABLE 5.10. Two-Way Classification (Without Replication) of Nitroglycerin Data

In using a machine computing technique, only the total, not the separate squares, would be tabulated for the next-to-last row and column.

Volume of sample	Operator			$\sum_j x_{ij}$	$(\sum_j x_{ij})^2$	$\sum_j x_{ij}^2$
	No. 1	No. 2	No. 3			
V_1	4	0	7	11	121	65
V_2	1	2	1	4	16	6
V_3	6	3	8	17	289	109
V_4	7	0	6	13	169	85
V_5	2	9	8	19	361	149
$\sum_i x_{ij}$	20	14	30	64	956	
$(\sum_i x_{ij})^2$	400	196	900	1,496		
$\sum_i x_{ij}^2$	106	94	214			414

Note: $c = 3; r = 5; n = 15.$

TABLE 5.11. Analysis of Variance for Nitroglycerin Data

Source of variation	Sum of squares	Degrees of freedom	Mean square
Among volumes	45.600	4	11.400
Among operators	26.133	2	13.067
Residual	69.200	8	8.650
Total	140.933	14	

$$\frac{15(414) - (64)^2}{15} = \frac{2,114}{15} = 140.933$$

$$\frac{5(956) - (64)^2}{15} = \frac{684}{15} = 45.600$$

$$\frac{3(1,496) - (64)^2}{15} = \frac{392}{15} = 26.133$$

$$\frac{2,114 - 392 - 684}{15} = \frac{1,038}{15} = 69.200$$

To test at the 5% significance level the hypothesis that the volume of the sample used in the determination has no effect on the result, we compare

$$F = \frac{11.400}{8.650} = 1.318$$

with $F_{.05}(4, 8) = 3.84$, from Appendix Table 5. We cannot conclude that the percentage of nitroglycerin reported depends on the volume of the sample.

To test the null hypothesis that the column means are equal; i.e., that the choice of operator has no effect on the percentage of nitroglycerin found in the sample, we compare

$$F = \frac{13.067}{8.650} = 1.511$$

with $F_{.05}(2, 8) = 4.46$ from Appendix Table 5. Since 1.511 does not exceed 4.46, the data are consistent with the null hypothesis that the choice of the operator does not affect the percentage of nitroglycerin detected.

Example (Type II Problem). Suppose the three operators in the example above were chosen at random from a large number of chemists, and the five volumes were selected at random from lists showing the volumes drawn for determinations during the past year. In this case we would be justified in analyzing the components of variance.

Referring to Tables 5.9 and 5.11, we find the estimates

$$cs_\alpha^2 + s_4^2 = 11.400$$

$$rs_\beta^2 + s_4^2 = 13.067$$

$$s_4^2 = 8.650$$

so that

$$s_\alpha^2 = 0.917 \quad \text{and} \quad s_\beta^2 = 0.883$$

The estimate of the total variance of a random observation is

$$0.883 + 0.917 + 8.650 = 10.450$$

We can use the F tests of the previous example (Type I problem) to test whether $\sigma_\alpha^2 = 0$ and whether $\sigma_\beta^2 = 0$. The variability caused by the operator and the variability caused by the volume are thus shown to be not significantly different from zero. Aside from the statistical significance (in the technical sense of the size of the effect relative to the random fluctuation), we see by inspection that the standard deviation of a random determination ($\sqrt{10.450} = 3.23$) is, in a practical sense, negligibly more than the standard deviation of a determination with operator and volume effects eliminated ($\sqrt{8.650} = 2.94$). The experimental data indicate that little increase in the reproducibility of determinations is achieved by holding either operator or volume fixed; however, the variance estimates are subject to considerable uncertainty because of the small size of the experiment.

5.5.2. Two-Factor Analysis With Replication

Only the Type I problem is treated here. Suppose that in Table 5.8 the set of treatments is applied not once, but m times under the same conditions. Then the cell, or treatment, (i, j) will contain the m observations x_{ij1}, $x_{ij2}, \ldots, x_{ijm}$. (Reference 5.13 treats the case in which the number of observations varies from cell to cell.) Here we have the mathematical model

$$x_{ijt} = \mu + \alpha_i + \beta_j + \delta_{ij} + e_{ijt} \qquad \begin{aligned} &(i = 1, 2, \cdots, r) \\ &(j = 1, 2, \cdots, c) \\ &(t = 1, 2, \cdots, m) \end{aligned}$$

where

μ = over-all mean

α_i = row main effect

β_j = column main effect

δ_{ij} = interaction of factor 1 at level i and factor 2 at level j (a joint effect, beyond the total of their individual effects)

e_{ijt} = random deviation from the mean position of the tth item receiving factor 1 at level i and factor 2 at level j. The e_{ijt} have independent normal distributions with mean zero and common variance σ_0^2.

In the Type I model, $\sum \alpha_i = \sum \beta_j = 0$, $\sum\limits_{j} \delta_{ij} = 0$ for $j = 1, 2, \ldots, c$, and $\sum\limits_{j} \delta_{ij} = 0$ for $i = 1, 2, \ldots, r$. The replication of the treatments makes it possible to estimate the interaction. The total number of observations n is rcm. The choice of successive items for each cell and the order of experimentation on them should be random among all rcm observations.

We begin the computations by carrying out the operations indicated in Tables 5.8 and 5.9 on the *sums* of the observations in each cell, with $n = rcm$. Let

$$X_{ij} = \sum_{t=1}^{m} x_{ijt}$$

The calculations on the X's yield the first four rows of the analysis of variance shown in Table 5.12. The remaining two rows are then computed, using the total sum of squares

$$\frac{n \sum\limits_{i,j,t}^{r,c,m} x_{ijt}^2 - \left(\sum\limits_{i,j,t}^{r,c,m} x_{ijt}\right)^2}{n}$$

As the first step in significance testing, we test the interaction against the within-treatments mean square. We compare

$$F = \frac{s_4^2}{s_6^2}$$

with $F_\alpha[(r-1)(c-1), rc(m-1)]$ from Appendix Table 5. If $F \leqslant F_\alpha$, we cannot reject the null hypothesis that the interactions are zero. If $F > F_\alpha$, we reject at the significance level α the null hypothesis that there is zero interaction, on the grounds that such a large value of F would occur with probability only α if the null hypothesis held.

Interaction can appear to be present for the following reasons:

1. The two factors operating together have effects beyond the combination of their separate effects; that is, a non-zero interaction is actually present. For instance, this might occur in the example given in Sec. 5.5.1 if one operator, though accurate when using large samples, tended to underestimate the percentage of nitroglycerin in small samples.

2. Significant interactions will appear with probability α as a result of random fluctuations alone.

TABLE 5.12. ANALYSIS OF VARIANCE FOR TWO-WAY CLASSIFICATION
(WITH REPLICATION)

$$x_{ij} = \sum_{l=1}^{m} x_{ijl}$$

Source of variation	Sum of squares	Degrees of freedom	Mean square
Among rows	$(2) = \dfrac{r\sum_i \left(\sum_j X_{ij}\right)^2 - \left(\sum_{i,j} X_{ij}\right)^2}{n}$	$r - 1$	$s_2^2 = \dfrac{(2)}{r-1}$
Among columns	$(3) = \dfrac{c\sum_j \left(\sum_i X_{ij}\right)^2 - \left(\sum_{i,j} X_{ij}\right)^2}{n}$	$c - 1$	$s_3^2 = \dfrac{(3)}{c-1}$
Interaction (residual for X's)	$(4) = (1) - (2) - (3)$	$(r-1)(c-1)$	$s_4^2 = \dfrac{(4)}{(r-1)(c-1)}$
Subtotal (total for X's)	$(1) = \dfrac{rc\sum_{i,j} X_{ij}^2 - \left(\sum_{i,j} X_{ij}\right)^2}{n}$	$rc - 1$	
Within treatments	$(6) = (5) - (1)$	$rc(m-1)$	$s_6^2 = \dfrac{(6)}{rc(m-1)}$
Total	$(5) = \dfrac{n\sum_{i,j,l} x_{ijl}^2 - \left(\sum_{i,j,l} x_{ijl}\right)^2}{n}$	$n - 1$	

3. Apparent interactions may be one of several unexplained effects occurring when important variables are left out of the analysis.

To test for row or column main effects, we form an F ratio with the row or column mean square in the numerator and a mean square that is an appropriate estimate of uncontrolled variability in the denominator. In an experiment designed with proper randomization, such an estimate is provided by the within-treatments mean square (s_6^2 in Table 5.12). Unfortunately, however, experiments may be performed without adequate randomization of the order in which treatments are applied, so that background conditions may remain nearly constant within each treatment, but vary from treatment to treatment. As a result, the uncontrolled variation in background conditions tends to enlarge the mean squares for interaction and main effects, but not the within-treatments mean square. The appropriate estimate of uncontrolled variability is, in this case, the interaction mean square s_4^2. (For further discussion related to the somewhat controversial question of which denominator to use, see Ref. 5.5, pp. 136–39; Ref. 5.12, pp. 337–49; and Ref. 5.15, pp. 119–23.)

When significant effects, especially with interactions, are found in an analysis of variance, it is helpful to present them graphically. (For a sample graph, see Fig. 5.1 in Sec. 5.6.) If interactions are significant, further analysis may be desirable (Ref. 5.4, pp. 139–46; Ref. 5.1, pp. 270–72; and Ref. 5.9, pp. 456–79).

Example. In the course of deciding on tactical uses for two rockets, an operations analyst requests the experiment reported in Table 5.13. Three types of planes are used five times with each of two types of rockets. The entries in Table 5.13 represent coded evaluations of the target destruction obtained in the tests. The experiment is designed to test whether the target destruction depends on the type of rocket or the type of plane, or on the interaction of the two factors, making use of significance tests at the 5% level.

To analyze the data, we replace the five values in each cell by the sum of those values, and carry out the preliminary calculations shown in Table 5.14. (In this table only the sums have been entered in the next-to-last row and column.) Table 5.15 shows the analysis of variance.

TABLE 5.13. Two-Way Classification (With Replication) of Rocket-Comparison Data

Rocket type	Plane type		
	I	II	III
A	0	1	6
	2	3	5
	3	3	7
	6	5	6
	5	6	6
B	3	2	6
	5	3	7
	7	4	5
	2	3	8
	5	6	8

TABLE 5.14. Calculations on Rocket-Comparison Data

Rocket type	Plane type			$\sum_j X_{ij}$	$(\sum_j X_{ij})^2$	$\sum_j X_{ij}^2$
	I	II	III			
A	16	18	30	64		1,480
B	22	18	34	74		1,964
$\sum_i X_{ij}$	38	36	64	138	9,572	
$(\sum_i X_{ij})^2$				6,836		
$\sum_i X_{ij}^2$	740	648	2,056			3,444

Test for interaction

$$F = \frac{0.933}{2.967} = 0.314$$

Since this F value is less than 1 and therefore certainly less than $F_{.05}(2, 24)$, the interaction is not significant. Incidentally, F values less than 1 occur with a substantial frequency in such computations if the null hypothesis is true.

Test for rocket effects

$$F = \frac{3.333}{2.967} = 1.12 \qquad \text{and} \qquad F_{.05}(1, 24) = 4.26$$

We conclude that the difference among rockets is not significant.

Test for plane effects

$$F = \frac{24.400}{2.967} = 8.22 \qquad \text{and} \qquad F_{.05}(2, 24) = 3.40$$

We conclude that the difference among planes is significant.

TABLE 5.15. ANALYSIS OF VARIANCE FOR ROCKET-COMPARISON DATA

Source of variability	Sum of squares	Degrees of freedom	Mean square
Among rockets	3.333	1	3.333
Among planes	48.800	2	24.400
Interaction	1.867	2	0.933
Subtotal	54.000	5	
Within treatments	71.200	24	2.967
Total	125.200	29	

$$\frac{6(3,444) - (138)^2}{30} = \frac{1,620}{30} = 54.000$$

$$\frac{2(9,572) - (138)^2}{30} = \frac{100}{30} = 3.333$$

$$\frac{3(6,836) - (138)^2}{30} = \frac{1,464}{30} = 48.800$$

$$\frac{1,620 - 1,464 - 100}{30} = \frac{56}{30} = 1.867$$

$$\frac{30(0^2 + 2^2 + 3^2 + 6^2 + 5^2 + 3^2 + 5^2 + \cdots + 8^2) - (138)^2}{30} = 125.200$$

$$125.200 - 54.000 = 71.200$$

5.6. Multiple-Factor Analysis

The methods described above can be extended to cover any number k of factors. We treat such data in k individual $(k-1)$-factor tables obtained by summing observations over each factor in turn. After the $(k-1)$-factor tables have been calculated separately, the residual for an experiment without replication is found by subtracting all sums of squares in the $(k-1)$-factor tables from the total sum of squares of deviations of individual observations from the over-all mean. The complexity and number of interactions increase rapidly, of course, but the method is general.

The following example shows the analysis of an experiment involving three factors.

Example. The observed variable (coded) is the outside diameter of a propellant grain. The purpose of the experiment is to study the effects on outside diameter of intentionally varying extrusion conditions. The factors are extrusion rate (factor A, three levels), die temperature (factor B, three levels), and charge temperature (factor C, two levels). The upper number of each pair of entries in Table 5.16 corresponds to charge temperature C_1; the lower, to charge temperature C_2.

First we compute 3 two-factor analyses on the sums of observations over each factor individually. Each of these analyses is obtained by considering the levels of one factor as replications and proceeding as

TABLE 5.16. Three-Way Classification (Without Replication) of Outside-Diameter Data

The upper number of each pair of entries corresponds to charge temperature C_1; the lower, to charge temperature C_2.

Die temp. B	Extrusion rate A		
	A_1	A_2	A_3
B_1	78	66	56
	34	23	5
B_2	85	84	75
	70	52	54
B_3	92	93	96
	79	70	64

in Sec. 5.5.2. In Table 5.17, for instance, the data for extrusion-rate versus die-temperature analysis were obtained by summing over the two levels of charge temperature for each entry. Table 5.18 shows the analysis of variance for this case.

Tables 5.18, 5.20, and 5.22 can be summarized, and the three-way analysis completed as in Table 5.23. The sums of squares are carried to several decimal places to provide sufficient significant figures in the relatively small residual sum of squares obtained by subtraction. An experienced variance analyst can omit intermediate Tables 5.17–5.22 by adjoining simple sums (but not sums of squares) to Table 5.16 and entering the final sums of squares directly into Table 5.23.

The F ratios for testing significance of the main effects and two-factor interactions are formed by dividing the appropriate mean squares by the residual mean square. Thus we have

	F ratio	*Critical value*	*Significance*
$F_A = \dfrac{324.667}{21.806} = 14.89$		$F_{.05}(2, 4) = 6.94$	significant
$F_B = \dfrac{2,340.667}{21.806} = 107.34$			significant
$F_C = \dfrac{4,170.889}{21.806} = 191.27$		$F_{.05}(1, 4) = 7.71$	significant
$F_{AB} = \dfrac{54.083}{21.806} = 2.48$		$F_{.05}(4, 4) = 6.39$	not significant
$F_{AC} = \dfrac{48.222}{21.806} = 2.21$			not significant
$F_{BC} = \dfrac{272.222}{21.806} = 12.48$			significant

All three factors thus have real main effects, but the effects of die temperature and charge temperature are not independent. Just what these effects are should be evaluated by referring to the original data in Table 5.16. The effects may then be presented graphically, as in Fig. 5.1. If interactions are significant, further analysis may be desirable (Ref. 5.4, pp. 139–46 and Ref. 5.1, pp. 270–72).

TABLE 5.17. Two-Way Table Derived From
Outside-Diameter Data

Die temp. B	Extrusion rate A			$\sum_j X_{ij}$	$(\sum_j X_{ij})^2$	$\sum_j X_{ij}^2$
	A_1	A_2	A_3			
B_1	112	89	61	262		24,186
B_2	155	136	129	420		59,162
B_3	171	163	160	494		81,410
$\sum_i X_{ij}$	438	388	350	1,176	489,080	
$(\sum_i X_{ij})^2$				464,888		
$\sum_i X_{ij}^2$	65,810	52,986	45,962			164,758

TABLE 5.18. Analysis of Variance for
Two-Way Table 5.17

Source of variation	Sum of squares	Degrees of freedom	Mean square
Extrusion rate A	649.333	2	324.667
Die temperature B	4,681.333	2	2,340.667
Interaction $A \times B$	216.333	4	54.083
Subtotal $A \times B$	5,547.000	8	

$$\frac{9\,(164,758) - (1,176)^2}{18} = \frac{99,846}{18} = 5,547.000$$

$$\frac{3\,(464,888) - (1,176)^2}{18} = \frac{11,688}{18} = 649.333$$

$$\frac{3\,(489,080) - (1,176)^2}{18} = \frac{84,264}{18} = 4,681.333$$

$$\frac{99,846 - 11,688 - 84,264}{18} = \frac{3,894}{18} = 216.333$$

If several observations had been taken for each of the 18 treatments, we would have had a within-treatments mean square and could have tested the significance of the three-factor interaction $A \times B \times C$ also. Our failure to replicate must be based on an assumption that the three-

TABLE 5.19. TWO-WAY TABLE DERIVED FROM
OUTSIDE-DIAMETER DATA

Charge temp. C	Extrusion rate A			Sum	Squared sum	Sum of squares
	A_1	A_2	A_3			
C_1	255	243	227	725		175,603
C_2	183	145	123	451		69,643
Sum	438	388	350	1,176	729,026	
Squared sum				464,888		
Sum of squares..	98,514	80,074	66,658			245,246

TABLE 5.20. ANALYSIS OF VARIANCE FOR
TWO-WAY TABLE 5.19

Source of variation	Sum of squares	Degrees of freedom	Mean square
Extrusion rate A	649.333	2	324.667
Charge temperature C ..	4,170.889	1	4,170.889
Interaction $A \times C$	96.444	2	48.222
Subtotal $A \times C$	4,916.667	5	

$$\frac{6\,(245{,}246) - (1{,}176)^2}{18} = \frac{88{,}500}{18} = 4{,}916.667$$

$$\frac{3\,(464{,}888) - (1{,}176)^2}{18} = \frac{11{,}688}{18} = 649.333$$

$$\frac{2\,(729{,}026) - (1{,}176)^2}{18} = \frac{75{,}076}{18} = 4{,}170.889$$

$$\frac{88{,}500 - 11{,}688 - 75{,}076}{18} = \frac{1{,}736}{18} = 96.444$$

factor interaction is negligible compared with the uncontrolled variability. The analysis of variance for the three-factor experiment with replication is obtained from the analysis without replication by the same extension used for two factors (Sec. 5.5.2); with proper randomization

TABLE 5.21. Two-Way Table Derived From
Outside-Diameter Data

Charge temp. C	Die temperature B			Sum	Squared sum	Sum of squares
	B_1	B_2	B_3			
C_1	200	244	281	725		178,497
C_2	62	176	213	451		80,189
Sum	262	420	494	1,176	729,026	
Squared sum				489,080		
Sum of squares..	43,844	90,512	124,330			258,686

TABLE 5.22. Analysis of Variance for
Two-Way Table 5.21

Source of variation	Sum of squares	Degrees of freedom	Mean square
Die temperature B	4,681.333	2	2,340.667
Charge temperature C ...	4,170.889	1	4,170.889
Interaction $B \times C$	544.444	2	272.222
Subtotal $B \times C$	9,396.667	5	

$$\frac{6(258,686) - (1,176)^2}{18} = \frac{169,140}{18} = 9,396.667$$

$$\frac{3(489,080) - (1,176)^2}{18} = \frac{84,264}{18} = 4,681.333$$

$$\frac{2(729,026) - (1,176)^2}{18} = \frac{75,076}{18} = 4,170.889$$

$$\frac{169,140 - 84,264 - 75,076}{18} = \frac{9,800}{18} = 544.444$$

of the order in which observations are made, the within-treatments
mean square is the appropriate denominator for all the F ratios. If, on
the other hand, the experiment were performed with replication but
without adequate randomization, then for testing main effects and two-
factor interactions, the appropriate denominator would be the three-

TABLE 5.23. THREE-WAY ANALYSIS OF VARIANCE
FOR OUTSIDE-DIAMETER DATA

Source of variation	Sum of squares	Degrees of freedom	Mean square
Extrusion rate A	649.333	2	324.667
Die temperature B ...	4,681.333	2	2,340.667
Charge temperature C .	4,170.889	1	4,170.889
Interaction $A \times B$ ...	216.333	4	54.083
Interaction $A \times C$ ...	96.444	2	48.222
Interaction $B \times C$ ...	544.444	2	272.222
Residual	87.224	4	21.806
Total	10,446.000	17	

$$\frac{18(78^2 + 34^2 + 85^2 + 70^2 + 92^2 + 79^2 + 66^2 + 23^2 + \cdots + 64^2) - (1,176)^2}{18}$$

$$= \frac{(18)(87,278) - (1,176)^2}{18} = 10,446.000$$

$$10,446.000 - (649.333 + 4,681.333 + 4,170.889 + 216.333 + 96.444 + 544.444)$$
$$= 87.224$$

factor interaction mean square. (See the similar discussion in Sec. 5.5.2, where references are given.)

The analysis of the corresponding Type II problem has a similar general form but differs in the formation of the F ratios; see the references in Sec. 5.5.2.

BIBLIOGRAPHY

5.1. Anderson, R. L., and T. A. Bancroft. Statistical Theory in Research. New York, McGraw-Hill, 1952.

5.2. Bartlett, M. S. "The Use of Transformations," BIOMETRICS, Vol. 3 (1947), pp. 39–52.

5.3. Cochran, W. G. "Some Consequences When the Assumptions for the Analysis of Variance Are Not Satisfied," BIOMETRICS, Vol. 3 (1947), pp. 22–38.

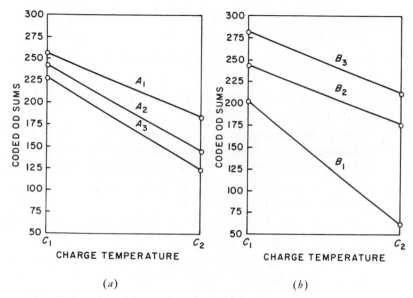

FIG. 5.1. Graphs of the Derived Two-Way Data of Tables 5.19 and 5.21, Showing the Non-Parallelism Caused by Interaction. (a) Coded outside-diameter sums versus charge temperature for three levels of extrusion rate A. The interaction $A \times C$ is not significant. (b) Coded outside-diameter sums versus charge temperature for three levels of die temperature B. The interaction $B \times C$ is significant.

5.4. Cochran, W. G., and G. M. Cox. Experimental Designs. New York, Wiley, 1950.

5.5. Dixon, W. J., and F. J. Massey, Jr. Introduction to Statistical Analysis. New York, McGraw-Hill, 1951. Chap. 10 on analysis of variance and Chap. 12 on analysis of covariance.

5.6. Eisenhart, C. "The Assumptions Underlying the Analysis of Variance," BIOMETRICS, Vol. 3 (1947), pp. 1–21.

5.7. Fisher, R. A., and F. Yates. Statistical Tables, 4th rev. ed. Edinburgh, Oliver & Boyd, Ltd., 1953.

5.8. Frankford Arsenal. Factorially Designed Experiments, by C. W. Churchman. Philadelphia, Frankford Arsenal, 13 March 1944. (Statistical Memorandum 1.)

5.9. Hald, A. Statistical Theory With Engineering Applications. New York, Wiley, 1952. Chap. 16 and 17.

5.10. Kempthorne, O. The Design and Analysis of Experiments. New York, Wiley, 1952.

5.11. Mann, H. B. Analysis and Design of Experiments. New York, Dover, 1949.

5.12. Mood, A. M. Introduction to the Theory of Statistics. New York, McGraw-Hill, 1950. Chap. 13 on regression and linear hypotheses and Chap. 14 on experimental designs and analysis of variance.

5.13. Rao, C. R. Advanced Statistical Methods in Biometric Research. New York, Wiley, 1952. Pp. 94–102.

5.14. Snedecor, G. W. Statistical Methods, 4th ed. Ames, Iowa, Iowa State College Press, 1946. Chap. 10–12.

5.15. Villars, D. S. Statistical Design and Analysis of Experiments. Dubuque, Iowa, William C. Brown Co., 1951.

5.16. Youden, W. J. Statistical Methods for Chemists. New York, Wiley, 1951. Chap. 6–10 treat the design and setup of experiments and the analysis of variance.

Chapter 6

FITTING A FUNCTION OF ONE OR MORE VARIABLES

6.0. Introduction

Often we want to predict the value of a variable y for any given values of one or more variables x_j. Sometimes a physical law connects the variables so that y may be expressed as a function of the x's. For instance, the formula $y = \frac{1}{2} gx^2$ enables us to predict the distance y traversed by a falling body in time x. However, the x and y values may follow a regular law only ideally, so that if they are plotted as points on an xy plane, they lie scattered about the curve that represents the ideal formula. Experimenters find such a "scatter diagram" valuable in exploratory work for relationships among variables. (For an example of a scatter diagram see Fig. 6.3 in Sec. 6.1.1.)

From past theoretical or experimental work it may be hypothesized that the relation among several variables is of a given form (e.g., a second-degree polynomial) without necessarily specifying the numerical values of all the constants in the equation. **Regression analysis** provides a systematic technique for estimating, with confidence limits, the unspecified constants from a new set of data, or for testing whether the new data are consistent with the hypothesis.

An important special case of a function of x's for predicting y is the linear function. Section 6.1 treats y as a linear function of one other variable x; Sec. 6.2 treats y as a linear function of several x's; Sec. 6.1.6 gives a test for linearity; Sec. 6.3 discusses transformations of non-linear data to reduce the problem to the linear case and, in particular, treats y as a polynomial function of x; and Sec. 6.4 considers the important statistical question of *planning* the experiment for fitting a function.

By means of *analysis of variance* (Chapter 5), we may determine whether certain chosen values of each of several variables $x_1, x_2, \cdots, x_k$ differ in their effects on a dependent variable y; the x_j need not be continuous variables, and their effects need not have any given functional form. By *regression analysis* we may determine the functional effect of continuous variables $x_1, x_2, \cdots, x_k$. Regression analysis can be considered a special case of variance analysis, but it is treated separately because of its special techniques and importance. A third method, the *analysis of covariance,* is especially important when some condition of the experiment cannot be held constant. By covariance analysis we can describe and (if desired) remove the effect of this changing condition, and test the effect of the other variables on the remaining variation in y. For analysis of covariance, see Ref. 6.4, Chap. 12; Ref. 6.13, Chap. 12; and Ref. 5.15, pp. 173–81.

A statistically significant regression relation of y on x (and similarly for more than one independent variable) is no indication that the "independent" variable x causes the observed change in the "dependent" variable y, as both may be caused to vary by a neglected third variable. For example, if the order is not random during an experiment (as when the successive x values are chosen in decreasing order), a change in y may reflect not the influence of x, but that of some hidden variable, such as temperature warm-up or tool wear, that changes steadily during the course of the experiment. If the identity of the basic causative variables is known from scientific or engineering knowledge of the phenomenon, regression analysis of appropriate data serves to estimate the constants of the relation (see Ref. 6.7, pp. 522–26, 613–14, and 634).

Though a regression equation is not necessarily built on a causal relationship, it does provide a valid prediction of the y value for each possible set of x values to within assessable random fluctuation. However, the prediction is no longer valid if the situation has changed since the data were taken.

The true or population **regression curve of y on x** gives for each x the mean of the corresponding y distribution. The equation or curve obtained as a result of data analysis is an approximation to the true regression curve and may be used to predict y values. In the linear case, the curve is called the "regression line." When there are several factors $x_1, x_2, \cdots, x_k$, the preceding statements hold if the words "surface" and "plane" are substituted for "curve" and "line."

Regression analysis may be applied to data gathered in two ways (Fig. 6.1):

1. The amount of the effect y may be measured for certain values of the factor x chosen *at will* by the experimenter (**Type I problem**).

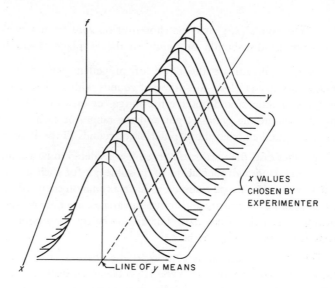

(*a*) Type I problem.

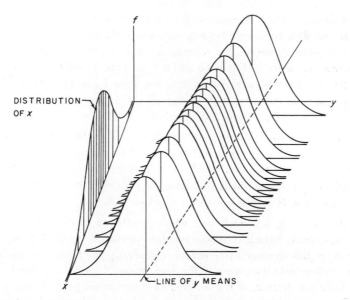

(*b*) Type II problem.
FIG. 6.1. Two Models for Regression Experiments.

2. The points (x, y) may be chosen *at random* from a two-dimensional normal distribution of points in the xy plane (**Type II problem**).

Thus, to investigate the effect of propellant-grain weight on rocket range, we may (1) load the different rounds with grains having particular weights of interest and observe the range for each one (Type I problem); or (2) measure the grain weight and the range for each of several rounds drawn at random from the population of rounds (Type II problem).

In either case, the y value observed is considered to be a random observation from a normal population of possible y's for each particular x used. These y populations are assumed to have a common standard deviation. The assumptions of normality and common variance are not necessary for simply fitting an equation from data, but are necessary for significance tests and confidence intervals given here.

The joint two-dimensional normal distribution from which the point (x, y) is drawn at random in Type II problems is illustrated in Fig. 6.2. Not only is each of x and y normally distributed by itself irrespective of the value of the other variable, but for each fixed x it can be shown that the distribution of y is normal, with its mean a linear function of x, and that all the y distributions have a common standard deviation. By the previously given definition, the regression curve of y on x is a straight line in this case, as shown in Fig. 6.1b and 6.2. In Type I problems the regression curve need not be straight; if it is, the Type I and Type II problems have so much in common that almost all the analysis is the same, and often the types have not been distinguished.

When a regression function has been derived from the data to represent the relationship between the effect and the given factor, we may test the reliability of that function, which is subject to the sampling error of the original data. It is in the investigation of reliability that the analyses for Type I and Type II regressions differ slightly.

As shown in Fig. 6.1, we have a whole population of y's for each x. All these populations have a common standard deviation, denoted by $\sigma_{y|x}$. If $\sigma_{y|x}$ is large, the regression line (or more generally, the regression surface), which predicts only the *mean* of the population of y's for a given x, will be of little use. Indeed, if the number of observations (x, y) is small, not even the mean will be estimated very precisely by the fitted regression equation.

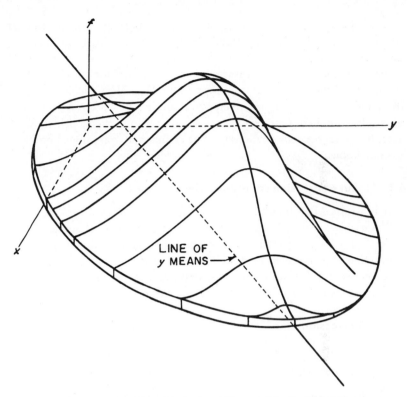

FIG. 6.2. Two-Dimensional Normal Distribution.

In a regression analysis the assumption is made that the mean of y is a linear function of the x's (more generally, a function of given form). If the resulting equation is used for predicting beyond the range of x's for which this assumption holds true, errors not accounted for in the confidence intervals for the regression will occur. For this reason, extrapolation is inadvisable.

The general method used in estimating a population regression curve from sample data is the **method of least squares**. The sample regression curve of y on x of given degree is the curve among all those of that degree that minimizes the sum of squares of vertical (y) deviations of the observed points from the curve. The method of least squares has general application in curve fitting (Sec. 6.3).

Not all the possible types of problems in fitting equations to data can be treated rigorously by the methods of this Manual; that is, as Type I or Type II problems. Models that are of neither type arise whenever the observations on x include measurement error in addition to any inherent variability. (The variation of y in Type I and Type II problems may result from measurement error or inherent variability, or both.) A statistician should be consulted if a problem does not appear to fit the models provided here; however, rigorous solutions to some of the more complicated problems are not known.

6.1. Linear Regression (Two Variables)

It is assumed that there are only two variables of interest, an "independent" variable x, and a "dependent" variable y, and that a sample $(x_1, y_1; x_2, y_2; \cdots; x_n, y_n)$ is drawn. The curve of the y mean values for the population is assumed to be a straight line. A test of this assumption is given in Sec. 6.1.6. If the curve is not straight, it may be known, or perhaps can be assumed, to be of a form that can be treated as explained in Sec. 6.3.

6.1.1. Regression Line

The equation of the **sample regression line of y on x** is

$$y' = a + bx$$

where the **regression coefficient**, the slope of the regression line, is

$$b = \frac{\sum (x - \bar{x})(y - \bar{y})}{\sum (x - \bar{x})^2} = \frac{n \sum xy - \sum x \sum y}{n \sum x^2 - (\sum x)^2}$$

and the y intercept is

$$a = \bar{y} - b\bar{x} = \frac{\sum y - b \sum x}{n}$$

(See Sec. 6.1.6 for formulas if there are several values of y for each value of x.) The general equation may also be written as

$$y' - \bar{y} = b(x - \bar{x})$$

which shows that the line passes through the center of gravity of the observed points. The prime (y') distinguishes a predicted or calculated value from an observed value.

The sample regression line given here is an estimate of the true or population regression line of y on x. The sample regression line is such that the sum of squares of vertical deviations of the observed points from this line is smaller than the corresponding sum for any other straight line; i.e., the sample regression line is a least-squares estimate of the unknown true line.

In a Type I problem, although the observations are made by observing y at values of x chosen by the experimenter, it may be desired to predict what value of x would yield a specified value of y on the average; this may be done by solving for x in the regression equation of y on x given above. In a Type II problem, on the other hand, when we know y and wish to predict x, the regression line of x on y is appropriate and would be determined mathematically by minimizing the sum of squared *horizontal* deviations. This line is different from the regression line of y on x (unless all points lie exactly on one line), since the latter minimizes the sum of squared *vertical* deviations. The formulas are completely analogous, and can be obtained by interchanging x and y in the formulas given above. More simply, the predicted variable can always be labeled y, as in this Manual.

For some purposes, a satisfactory line can be fitted to the scatter diagram by eye, without computations.

Example. Fit a regression line of y on x to the data (Type II) of Table 6.1. This table gives 22 paired values of % nitrocellulose and Young's modulus in thousands of psi.

$$\sum x = 1{,}260.21 \qquad \sum x^2 = 72{,}706.0739$$

$$\sum y = 1{,}832.9 \qquad \sum y^2 = 187{,}657.71$$

$$n = 22 \qquad \sum xy = 108{,}805.243$$

A partial check of the sums of squares and cross-products is given by the relation

$$\sum (x - y)^2 = \sum x^2 - 2 \sum xy + \sum y^2$$

where the two sides of the equation are computed independently.

$$42{,}753.2979 \overset{?}{=} 72{,}706.0739 - 2\,(108{,}805.243) + 187{,}657.71$$

$$= 42{,}753.2979$$

TABLE 6.1. DATA FOR LINEAR-REGRESSION EXAMPLE

Propellant lot	Nitrocellulose, % x	Young's modulus at 70°F, in thousands of psi y
1	55.77	83.9
2	55.05	66.4
3	54.27	73.1
4	50.63	66.7
5	49.86	30.1
6	53.04	36.2
7	51.33	22.8
8	56.70	66.5
9	55.07	37.0
10	55.76	58.0
11	54.40	71.9
12	55.39	83.1
13	57.49	66.2
14	57.56	72.3
15	58.76	65.8
16	59.32	123.5
17	57.21	116.8
18	68.55	160.1
19	65.04	158.2
20	66.98	152.2
21	63.69	134.8
22	58.34	87.3

The regression coefficients are

$$b = \frac{83,876.437}{11,404.3817} = 7.35476$$

$$a = \frac{1,832.9 - (7.35476)(1,260.21)}{22} = -337.983$$

Therefore, the equation of the sample regression line of y on x is

$$y' = -337.98 + 7.3548x$$

This line is plotted as the central solid line on the scatter diagram of Fig. 6.3. (Other curves shown in this figure will be described in Sec. 6.1.2 and 6.1.4e.) To check the values for a and b, substitute them in the equation

$$\sum xy = a \sum x + b \sum x^2$$

which should hold, except for rounding error.

$$108,805.243 \overset{?}{=} (-337.983)(1,260.21) + (7.35476)(72,706.0739)$$

$$\cong 108,806.168$$

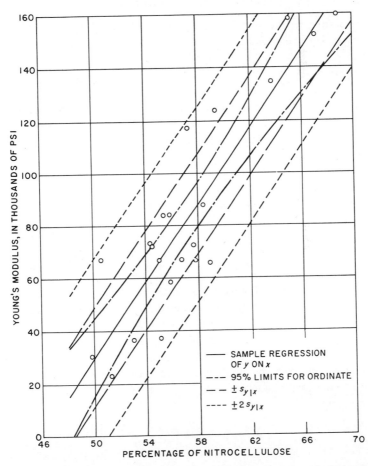

FIG. 6.3. Graph for Table 6.1, Showing Regression Line, $\pm s_{y|x}$ and $\pm 2s_{y|x}$ Limits for Individual Observation (Sec. 6.1.2), and 95% Confidence Limits for Regression Ordinate (Sec. 6.1.4e).

To illustrate the use of the regression line in predicting values of y, we compute y', given $x = 55.77\%$.

$$y' = -337.98 + (7.3548)(55.77) = 72.20 \times 10^3 \text{ psi}$$

The observed value of y is 83.9×10^3 psi.

6.1.2. Variation About the Regression Line

The scatter, in the vertical (y) direction, of the observed points about the regression line is measured by $s_{y|x}$, where

$$s_{y|x}^2 = \frac{\sum (y_i - y_i')^2}{n - 2} = \frac{n - 1}{n - 2}(s_y^2 - b^2 s_x^2) = \frac{n - 1}{n - 2} s_y^2 (1 - r^2)$$

Either of the latter two forms is to be used for computing. (Section 3.1 shows how to compute s_x^2 and s_y^2; r, the sample correlation coefficient, is defined in Sec. 6.1.3.) The quantity $s_{y|x}^2$ estimates that part of the variance of y left unexplained by the regression of y on x. It is defined with $n - 2$ rather than $n - 1$ in the denominator (Sec. 1.1.12) because two degrees of freedom are absorbed by the estimates a and b. The positive square root, $s_{y|x}$, is sometimes called the **standard error of estimate**.

The sum of squares of deviations of the y_i from their over-all mean $\bar{y}$ can be separated into two parts

$$\sum (y_i - \bar{y})^2 = \sum (y_i - y_i')^2 + \sum (y_i' - \bar{y})^2$$

That is, the total sum of squares equals the sum of squares of the deviations from the regression line plus the sum of squares of the deviations of the regression values from $\bar{y}$. This separation into two parts is similar to that described in the analysis of variance in Chapter 5. If we call $s_{y'}^2 = \sum (y_i' - \bar{y})^2/(n - 1)$ the variance accounted for by the regression, then (neglecting only the slight difference between $n - 1$ and $n - 2$) we can write the total variance s_y^2, like the total sum of squares, as the sum of two parts

$$s_y^2 \cong s_{y|x}^2 + s_{y'}^2.$$

In a Type II model, a similar relation holds exactly for the corresponding population variances

$$\sigma_y^2 = \sigma_{y|x}^2 + \sigma_{y'}^2.$$

This separation into parts is discussed further in Sec. 6.1.3.

If it is assumed that the y populations for the various x's have normal distributions about the true regression line with common standard deviation $\sigma_{y|x}$, then 68% of the combined populations lie within $\pm\sigma_{y|x}$ of the true regression line and 95% within $\pm 2\sigma_{y|x}$. With a large sample we expect to approximate this situation, taking the sample regression line and $s_{y|x}$ as good approximations to the true regression line and $\sigma_{y|x}$, respectively. For a more precise description of the distribution of the population about the regression line, see Sec. 6.1.4f; see also Ref. 6.14 on tolerance intervals for the regression line. (Tolerance intervals are defined in Sec. 1.1.11.)

Example. For the data of Table 6.1, we calculate

$$s_x^2 = \frac{n \sum x^2 - (\sum x)^2}{n(n-1)} = 24.6848$$

$$s_y^2 = \frac{n \sum y^2 - (\sum y)^2}{n(n-1)} = 1{,}664.3879$$

$$s_{y|x}^2 = \frac{n-1}{n-2}\left(s_y^2 - b^2 s_x^2\right) = 1.05(329.126) = 345.582$$

$$s_{y|x} = 18.59 \times 10^3 \, \text{psi} \qquad \text{and} \qquad s_y = 40.80 \times 10^3 \, \text{psi}$$

Lines are drawn on Fig. 6.3 at vertical distances of $\pm s_{y|x}$ and $\pm 2s_{y|x}$ units from the regression line. We see that 16 of the points (73%) lie within $\pm s_{y|x}$ of the regression line, and that all 22 points (100%) lie within $\pm 2s_{y|x}$.

6.1.3. Fraction of y Variance Accounted for by Regression—Correlation Coefficient r

From Sec. 6.1.2 it follows that the fraction of the population y variance accounted for by the regression on x is

$$\frac{\sigma_{y'}^2}{\sigma_y^2} = 1 - \frac{\sigma_{y|x}^2}{\sigma_y^2}$$

The square root of this fraction is the **population correlation coefficient** ρ; the sign is taken as positive or negative to agree with the sign of the slope of the population regression line. (A negative correlation indicates that high values of y are associated with low values of x.) The best estimate of ρ from the sample is the **sample correlation coefficient**

$$r = \frac{\sum (x - \bar{x})(y - \bar{y})}{(n - 1) s_x s_y}$$

$$= \frac{n \sum xy - \sum x \sum y}{\sqrt{[n \sum x^2 - (\sum x)^2][n \sum y^2 - (\sum y)^2]}}$$

If the regression coefficient has already been calculated, we can use the form

$$r = b \frac{s_x}{s_y}$$

The correlation coefficient always lies between -1 and $+1$. If, and only if, all points lie on the regression line, then $r = \pm 1$. If $r = 0$, the regression does not explain anything about the variation of y, and the regression line is horizontal ($y' = \bar{y}$).

The correlation coefficient is defined primarily for Type II problems, where r is an estimate of the true correlation coefficient ρ of the joint two-dimensional normal distribution (Fig. 6.2), irrespective of the random set of x's obtained. The correlation coefficient does not have such an interpretation in Type I problems, where the regression coefficient b is ordinarily used; however, it is formally correct and sometimes convenient to use the notation r and some of the associated formulas.

Example. For the data of Table 6.1, we calculate

$$r^2 = \frac{(n \sum xy - \sum x \sum y)^2}{[n \sum x^2 - (\sum x)^2][n \sum y^2 - (\sum y)^2]}$$

$$= \frac{(83,876.437)^2}{(11,404.3817)(768,947.21)} = 0.802254$$

$$r = 0.8957$$

We see from r^2 that the regression of y on x accounts for about 80% of the variance of y.

6.1.4. Reliability of Regression Measures

The significance tests and confidence intervals given below are exact only under the normality assumptions discussed in Sec. 6.0. With the exception of Subsec. b, all of Sec. 6.1.4 is applicable to Type I problems, but the test in Subsec. a is identical with a special case of Subsec. c, so that only Subsec. c to f need be considered in Type I problems. All of this section is applicable to Type II problems, Subsec. a again being equivalent to a special case of Subsec. c. (The tests of hypotheses in Subsec. c and d have somewhat different operating characteristic (OC) curves in Type I and Type II problems, but the difference is not a practical concern. See Sec. 1.1.8–1.1.9 for OC curves.)

a. Significance of Sample Correlation Coefficient r. Appendix Table 7 gives 1 and 5% critical points for the absolute value of r, denoted by $|r|$. We read the critical value under two variables and opposite $n - 2$ degrees of freedom. If the computed $|r|$ exceeds the critical value, we reject at that level of significance the null hypothesis that the population of x's and y's has zero correlation.

Example. For the data of Table 6.1, $r = 0.896$. Suppose we wish to test the hypothesis that there is zero correlation between the % nitrocellulose and Young's modulus at the 5% level of significance. From Appendix Table 7, under two variables and opposite $22 - 2 = 20$ degrees of freedom, we read 0.423 for the 5% critical point. Since 0.896 exceeds 0.423, we reject the null hypothesis. As explained in Sec. 1.1.7, the hypothesis is rejected because if it were true such a high value of $|r|$ would occur with probability less than 5%. The test is an equal-tails test, appropriate when we are interested in either positive or negative correlation, as is usually the case.

b. Confidence Interval for Population Correlation Coefficient ρ. This subsection applies only to Type II problems. With the help of Appendix Chart XI we can construct, on the basis of the correlation coefficient r from a sample of size n, a confidence interval for the true (population) correlation coefficient ρ. In 95% of our experiments the interval so constructed will contain ρ.

Example. For the data of Table 6.1, $r = 0.896$. Construct a 95% confidence interval for ρ.

Entering Appendix Chart XI with this abscissa, we obtain for sample size 22 (interpolating between curves labeled 20 and 25) the 95% confidence interval 0.75 to 0.96.

c. Significance Test and Confidence Interval for the Slope b of the Regression Line. The standard error of the regression coefficient b can be estimated from the sample as

$$s_b = \frac{s_{y|x}}{s_x\sqrt{n-1}}$$

where

$$s_x^2 = \frac{1}{n(n-1)}\left[n\sum x^2 - \left(\sum x\right)^2\right]$$

and

$$s_{y|x}^2 = \frac{n-1}{n-2}\left(s_y^2 - b^2 s_x^2\right)$$

To test the null hypothesis that the slope β of the true or population regression line has any stated value, say B, compute

$$t = \frac{b-B}{s_b}$$

Reject the null hypothesis at the significance level α if $|t|$ exceeds the critical value $t_{\alpha/2,n-2}$ given in Appendix Table 3 for $P(t) = \alpha/2$ and $f = n-2$ degrees of freedom (Ref. 6.2, pp. 402–3 and 548–51).

A $100(1-\alpha)\%$ confidence interval for the true slope β is

$$b \pm t_{\alpha/2,n-2}s_b$$

Example. For the data of Table 6.1, test the null hypothesis that β has the value $B = 0$.

We compute

$$s_b = \frac{s_{y|x}}{s_x\sqrt{n-1}} = \frac{18.5898}{\sqrt{(24.6848)(21)}} = 0.81649$$

and

$$t = \frac{b}{s_b} = \frac{7.35476}{0.81649} = 9.008$$

Since Appendix Table 3 gives $t_{.025,20} = 2.086$, we find the computed t, and hence b, significantly different from zero at the 5% level of significance.

To find a 95% confidence interval for the true slope β, we compute

$$b \pm t_{.025,20}\, s_b = 7.35476 \pm (2.086)\,(0.81649)$$

We find a 95% confidence interval of 5.652 to 9.058.

d. Significance of the Difference Between the Regression Coefficients b_1 and b_2 of Two Separate Equations. If two lines $y_1' = a_1 + b_1 x$ and $y_2' = a_2 + b_2 x$ have been determined from two sets of data with comparable variables, the t test can be used to test the significance of the difference between the slopes, under the assumption that the population standard errors of estimate for the two relations are equal. (This assumption can be tested as in Sec. 3.3 by forming the F ratio of the two values of $s_{y|x}^2$, say $s_{y_1|x}^2$ and $s_{y_2|x}^2$, and comparing it with the critical value in Appendix Table 5 for $n_1 - 2$ and $n_2 - 2$ degrees of freedom.)

We compute the best estimate $s_{y|x}'^2$ of the common variance $\sigma_{y\cdot x}^2$

$$s_{y|x}'^2 = \frac{(n_1 - 2)\, s_{y_1|x}^2 + (n_2 - 2)\, s_{y_2|x}^2}{n_1 + n_2 - 4}$$

Based on the pooled information from both samples, the estimate of the variance of the regression coefficient b_i is

$$s_{b_i}'^2 = \frac{s_{y|x}'^2}{(n_i - 1)\, s_{x_i}^2} \qquad\qquad (i = 1, 2)$$

and the standard deviation of the difference $b_1 - b_2$ is estimated from

$$s_{b_1-b_2}^2 = s_{b_1}'^2 + s_{b_2}'^2 = s_{y|x}'^2 \left[\frac{1}{(n_1 - 1)\, s_{x_1}^2} + \frac{1}{(n_2 - 1)\, s_{x_2}^2} \right]$$

To test the null hypothesis that the difference $\beta_1 - \beta_2$ between the true (population) regression coefficients has any stated value, say Δ, compute

$$t = \frac{b_1 - b_2 - \Delta}{s_{b_1-b_2}}$$

Reject the hypothesis at the significance level α if $|t|$ exceeds the critical value $t_{\alpha/2,\, n_1+n_2-4}$ given in Appendix Table 3.

For the comparison of more than two regression coefficients see Ref. 6.4, Chap. 12, or Ref. 6.7, Sec. 18.9.

e. Confidence Interval for an Ordinate to the True Regression Line. The ordinate of the sample regression line for any given $x = X$ (which need not be any of the observed x's) is calculated as

$$y' = a + bX$$

This y' differs from the true or population *mean* ordinate at $x = X$ which would be obtained if infinitely many observations could be made with the same value X of x. We can show how good our estimate y' of the true *mean* ordinate is by calculating $100(1 - \alpha)\%$ confidence limits

$$y' \pm t_{\alpha/2, n-2} \, s_{y|x} \sqrt{\frac{1}{n} + \frac{(X - \bar{x})^2}{(n - 1) s_x^2}}$$

where $t_{\alpha/2, n-2}$ is obtained from Appendix Table 3. In this way we can construct a confidence interval for any particular ordinate of interest. Notice that if X is set equal to $\bar{x}$ in the equation $y' = a + bX$, we have $y' = \bar{y}$. If $X = 0$, then $y' = a$, the y intercept.

To make an equal-tails test of the null hypothesis that the ordinate to the true regression line at the chosen X has any stated value Y, we compute

$$t = \frac{y' - Y}{s_{y|x} \sqrt{\dfrac{1}{n} + \dfrac{(X - \bar{x})^2}{(n - 1) s_x^2}}}$$

We reject the hypothesis at the significance level α if $|t|$ exceeds the critical value $t_{\alpha/2, n-2}$ given in Appendix Table 3.

Example. Using the data of Table 6.1, calculate a 95% confidence interval for the ordinate to the regression line of Fig. 6.3 for the percentage $x = X$ of nitrocellulose.

We compute

$$y' \pm t_{\alpha/2, n-2} \, s_{y|x} \sqrt{\frac{1}{n} + \frac{(X - \bar{x})^2}{(n - 1) s_x^2}}$$

with $n = 22$, $\bar{x} = 57.28$, $s_{y|x} = 18.59$, $(n - 1) s_x^2 = 518.38$, and $y' = -337.98 + 7.3548X$. From Appendix Table 3, we find $t_{.025, 20} = 2.086$. Hence the 95% confidence limits for the ordinate are

$$-337.98 + 7.3548X \pm (2.086)(18.59)\sqrt{\frac{1}{22} + \frac{(X - 57.28)^2}{518.38}}$$

$$= -337.98 + 7.3548X \pm \sqrt{68.35 + 2.901(X - 57.28)^2}$$

These limits are the curved lines plotted in Fig. 6.3. Note that the error in estimating the true mean ordinate is smallest at $X = \bar{x}$, and increases steadily as $|X - \bar{x}|$ increases.

f. Prediction Interval for Individual Value of y. For any given x the individual values of y are scattered above and below both the true and the sample regression lines. In practice, it may more often be of interest to know how closely one can predict an individual value of y rather than the mean value given by the regression line (Sec. 6.1.4e). The individual value of y corresponding to a given $x = X$ is, as in Sec. 6.1.4e, calculated as

$$y' = a + bX$$

The formula for a $100(1 - \alpha)\%$ **prediction interval** for y (explained below) is

$$y' \pm t_{\alpha/2, n-2} \, s_{y|x} \sqrt{1 + \frac{1}{n} + \frac{(X - \bar{x})^2}{(n - 1) s_x^2}}$$

To see what such an interval means, suppose that we conduct an experiment of n observations and construct an interval for the value of y corresponding to some given $x = X$; and suppose that we take a single further observation y_{n+1} at $x = X$. The length of the prediction interval is determined so that in many repetitions of this whole procedure, on the average $100(1 - \alpha)\%$ of the intervals will contain the corresponding additional observation y_{n+1}. (See Ref. 5.12.)

Example. For the data of the example in Sec. 6.1.4e, the 95% prediction interval for a single observation y at $x = X = 60$ is

$$-337.98 + 7.3548X \pm 38.78 \sqrt{1 + \frac{1}{22} + \frac{(X - 57.28)^2}{518.38}}$$

$$= -337.98 + 7.3548X \pm \sqrt{1572.2 + 2.901(X - 57.28)^2}$$

$$= 103.31 \pm \sqrt{1593.7} = 103.31 \pm 39.92$$

$$= 63.39 \text{ and } 143.23$$

Evidently a single value of y (Young's modulus) can be predicted but poorly; this is because of intrinsic variability from sample to sample rather than lack of knowledge of the true regression line, as shown by the relative sizes of the square-root quantities here and in Sec. 6.1.4e.

6.1.5. Summary of Computations for Two-Variable Linear Regression

We show below the calculations for the data of Table 6.1. Accumulated multiplication on a desk computer should be used to find sums and sums of squares simultaneously.

1. *Sums and variances*

$$\sum x = 1{,}260.21 \qquad \sum x^2 = 72{,}706.0739$$

$$\sum y = 1{,}832.9 \qquad \sum y^2 = 187{,}657.71$$

$$n = 22 \qquad \sum xy = 108{,}805.243$$

$$s_x^2 = \frac{n \sum x^2 - (\sum x)^2}{n(n-1)} = \frac{11{,}404.3817}{(22)(21)} = 24.6848$$

$$s_y^2 = \frac{n \sum y^2 - (\sum y)^2}{n(n-1)} = \frac{768{,}947.21}{(22)(21)} = 1{,}664.3879$$

2. *Regression line*

$$y' = a + bx$$

$$b = \frac{n \sum xy - \sum x \sum y}{n \sum x^2 - (\sum x)^2} = \frac{83{,}876.437}{11{,}404.3817} = 7.35476$$

$$a = \bar{y} - b\bar{x} = \frac{\sum y - b \sum x}{n} = \frac{1{,}832.9 - (7.35476)(1{,}260.21)}{22} = -337.98$$

3. *Correlation coefficient*

$$r^2 = \frac{[n \sum xy - (\sum x)(\sum y)]^2}{[n \sum x^2 - (\sum x)^2][n \sum y^2 - (\sum y)^2]}$$

$$= \frac{(83{,}876.437)^2}{(11{,}404.3817)(768.947.21)} = 0.802254$$

$$r = \sqrt{0.802254} = 0.8957$$

4. *Standard error of estimate*

$$s_{y|x}^2 = \frac{n-1}{n-2}(s_y^2 - b^2 s_x^2) = 1.05\,(329.126) = 345.582$$

$$s_{y|x} = \sqrt{345.582} = 18.5898$$

5. *Standard error of regression coefficient b*

$$s_b = \frac{s_{y|x}}{s_x\sqrt{n-1}} = \frac{18.5898}{\sqrt{(24.6848)(21)}} = 0.81649$$

6. *Significance of r.* Compare $|r|$ with the critical value in Appendix Table 7 for 2 variables and $n-2$ degrees of freedom.

7. *Significance of b.* Compare $|t| = |b|/s_b$ with the critical value in Appendix Table 3 for $n-2$ degrees of freedom.

For further information on the calculations given above, see Sec. 6.1.4.

6.1.6. Test for Linearity of Regression

If the regression function of y on x is known to be a polynomial, the tests given in Sec. 6.3.1 and 6.3.2 can be used to determine the degree of that polynomial. In the latter case the x values used must be equally spaced. If the degree of the polynomial is one, the relation is linear.

A cruder test of linearity can be made by considering the sequence of signs of the deviations $y_i - y_i'$ from the fitted regression line in order of increasing x_i, and applying the run test for randomness described in Sec. 4.1. For example, in Fig. 6.3 the sequence of signs of $y_i - y_i'$ is

$$+ + - - + + - - + - + - + - - - - - + + + - -$$

with 10 $+$'s, 12 $-$'s, and 12 runs. If the population regression curve is non-linear, the number of runs usually tends to be less than if it were linear. For a test of linearity at the 5% significance level we compare 12 with the critical value $v_{.95} = 7$ in Appendix Table 10 for $n_1 = 10$ and $n_2 = 12$. Thus we do not reject the hypothesis of linearity. The same test can be applied to any fitted curve, not merely a straight line.

A much better test of linearity, or of any form of curve, is available in a Type I model if more than one observation of y is made for each selected value of x. An F test can be made of the variance about the regression

relative to the variance within groups, similar to the methods of Chapter 5.
Let the n_i values of y for each chosen x value x_i be called y_{ij}, where
$j = 1, 2, \cdot \cdot \cdot , n_i$ and $i = 1, 2, \cdot \cdot \cdot , m > 2$.

To test for a general linear relation between x and y

$$y = \alpha + \beta x$$

where the values of α and β are not specified, calculate the m means

$$\bar{y}_i = \frac{1}{n_i} \sum_{j=1}^{n_i} y_{ij} \qquad\qquad (i = 1, 2, \cdot \cdot \cdot , m)$$

and the m estimates of variance

$$s_{1i}^2 = \frac{1}{n_i(n_i - 1)} [n_i \sum_{j=1}^{n_i} y_{ij}^2 - (\sum_{j=1}^{n_i} y_{ij})^2] \qquad (i = 1, 2, \cdot \cdot \cdot , m)$$

These s_{1i}^2 estimate the variances of the y_{ij}'s for each x_i. In the regression
model these variances are all assumed equal (to $\sigma_{y|x}^2$). This assumption can
be tested as in Sec. 3.4. The m estimates of variance s_{1i}^2 can be combined
to give an estimate of the common variance $\sigma_{y|x}^2$

$$s_1^2 = \frac{\sum (n_i - 1) s_{1i}^2}{N - m}$$

with $N - m$ degrees of freedom, where $N = \sum n_i$. Here, and throughout
the following summations, the index of summation is i and the range (from
1 to m) will be omitted for the sake of simplicity.

The constants α and β are estimated by the least-squares line

$$y' = a + bx$$

fitted to the data, where

$$b = \frac{N \sum n_i x_i \bar{y}_i - (\sum n_i x_i)(\sum n_i \bar{y}_i)}{N \sum n_i x_i^2 - (\sum n_i x_i)^2}$$

and

$$a = \frac{\sum n_i \bar{y}_i - b \sum n_i x_i}{N}$$

(These follow directly from Sec. 6.1.1 by denoting the equal values of x
by the same symbol.) Using the values y' given by this least-squares line,

we form

$$s_2^2 = \frac{\sum n_i (\bar{y}_i - y_i')^2}{m - 2}$$

$$= \frac{N \sum n_i \bar{y}_i^2 - (\sum n_i \bar{y}_i)^2 - b [N \sum n_i x_i \bar{y}_i - (\sum n_i x_i)(\sum n_i \bar{y}_i)]}{N(m - 2)}$$

with $m - 2$ degrees of freedom. The latter form is preferred for computing. *If the hypothesis of linearity is true*, s_2^2 also estimates $\sigma_{y|x}^2$; *if not*, s_2^2 tends to exceed $\sigma_{y|x}^2$. To test the hypothesis of linearity, compute the F ratio

$$F = \frac{s_2^2}{s_1^2}$$

Reject the hypothesis at the 5% significance level if the calculated F is larger than the critical value $F_{.05}(m - 2, N - m)$ in Appendix Table 5.

If the hypothesis of linearity is *not rejected*, the significance tests and confidence intervals for b and any particular y' (such as a) can be made as in Sec. 6.1.4 on ungrouped data, except that $s_{y|x}^2$ is replaced by s_1^2 and the number of degrees of freedom $n - 2$ is replaced by $N - m$.

If the hypothesis of linearity is *rejected*, then some other less simple form of relation must be considered. The data may be used to discover such a relation, but rigorous statistical tests of the validity of the relation suggested by the data cannot be made with the same data. An entirely new set must be gathered.

See Ref. 6.7 (Sec. 18.3, 18.4, and 18.7) for a more extensive discussion of linearity tests.

The test has much more general application than that described above. It can be used with only slight changes to test *any* form of relation in which y' is linear in the constants to be determined; e.g., for testing a polynomial in x. The only changes are the following: (1) the equation of interest, rather than a straight line, is fitted by least squares as in Sec. 6.2 or 6.3; and (2) if the number of constants to be determined is $k + 1$ rather than 2 (where $k + 1 < m$), then $m - k - 1$ replaces $m - 2$ as the number of degrees of freedom in the denominator of the first expression for s_2^2, and the second expression for s_2^2 does not apply.

6.2. Multiple Linear Regression

If the variable y depends linearly on several factors, it is best to use a formula involving all of them for predicting y. Denote the k such factors, or variables, by $x_1, x_2, \cdots, x_k$. A sample of n observations would consist of n sets of values: $x_{11}, x_{21}, \cdots, x_{k1}, y_1; x_{12}, x_{22}, \cdots, x_{k2}, y_2; \cdots$; and $x_{1n}, x_{2n}, \cdots, x_{kn}, y_n$. The subscript denoting the particular observation is usually omitted, and the representative factor is denoted by x_j where $j = 1, 2, \cdots, k$.

The general discussion of the meaning of the regression equation in Sec. 6.0 and 6.1 applies to multiple as well as simple regression. If the whole population of values of $(x_1, x_2, \cdots, x_k, y)$ were known, the ordinate to the true or population regression plane at the x point $(x_1, x_2, \cdots, x_k)$ would be the y mean for that x point (Fig. 6.4). The values of $x_1, x_2, \cdots, x_k$ may be specified by the experimenter or they may occur

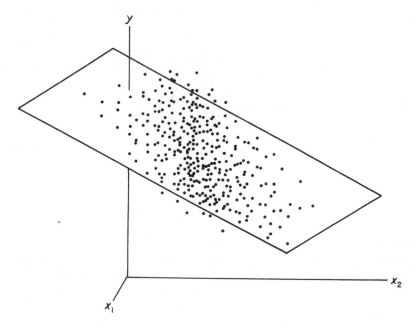

FIG. 6.4. Regression Plane.

randomly, giving rise to the Type I and Type II problems mentioned in Sec. 6.0. Intermediate mixed types may occur. The restrictions on Type I and Type II models in the multiple-regression case are analogous to those in the simple-regression case. In any case, each sample value of y is assumed to be a *random* observation from a normal population. These populations have a common standard deviation, but different means, depending linearly on $x_1, x_2, \cdots, x_k$.

The methods of this section presuppose a linear relationship between the mean value of y and the other variables. A rough check on this assumption can be made by graphing y against the x's taken one at a time (see Ref. 6.5, Chap. 11, 14, and 16). A general test of a regression surface, which requires several observations of y for each point $x_1, x_2, \cdots, x_k$, is available in Sec. 6.1.6.

If the regression curves seem to have a form other than linear, a transformation of variables may yield a linear relationship which would make the methods of this section applicable. However, if the forms of the curves are discovered from the data under analysis, they represent a hypothesis rather than an established fact and should be tested with a further experiment.

6.2.1. Multiple Linear Equation

The predicted value y' is expressed by the **sample regression equation**

$$y' = a + b_1 x_1 + b_2 x_2 + \cdots + b_k x_k$$

where the x_j are the variables affecting y, the **partial regression coefficients** b_j are determined as shown in Sec. 6.2.2, and

$$a = \bar{y} - b_1 \bar{x}_1 - b_2 \bar{x}_2 - \cdots - b_k \bar{x}_k$$

the sample averages being $\bar{y}, \bar{x}_1, \bar{x}_2, \cdots, \bar{x}_k$. Thus the regression equation could also be written

$$y' - \bar{y} = b_1(x_1 - \bar{x}_1) + b_2(x_2 - \bar{x}_2) + \cdots + b_k(x_k - \bar{x}_k)$$

This equation represents the best-fitting plane for the sample data plotted in $k + 1$ dimensions, in the sense of the least sum of squares of y deviations.

6.2.2. Determination of Partial Regression Coefficients

We determine the partial regression coefficient b_j in the regression equation by solving the *normal equations*

$$b_1 \sum (x_1 - \bar{x}_1)^2 + b_2 \sum (x_1 - \bar{x}_1)(x_2 - \bar{x}_2) + \cdots + b_k \sum (x_1 - \bar{x}_1)(x_k - \bar{x}_k) = \sum (x_1 - \bar{x}_1)(y - \bar{y})$$

$$b_1 \sum (x_2 - \bar{x}_2)(x_1 - \bar{x}_1) + b_2 \sum (x_2 - \bar{x}_2)^2 + \cdots + b_k \sum (x_2 - \bar{x}_2)(x_k - \bar{x}_k) = \sum (x_2 - \bar{x}_2)(y - \bar{y})$$

. . . .

$$b_1 \sum (x_k - \bar{x}_k)(x_1 - \bar{x}_1) + b_2 \sum (x_k - \bar{x}_k)(x_2 - \bar{x}_2) + \cdots + b_k \sum (x_k - \bar{x}_k)^2 = \sum (x_k - \bar{x}_k)(y - \bar{y})$$

Each sum has n terms for the n sample members. Thus

$$\sum (x_1 - \bar{x}_1)^2 = \sum_{i=1}^{n} (x_{1i} - \bar{x}_1)^2$$

but the second subscript has been omitted from these equations for convenience.

To simplify the notation, we let

$$a_{hj} = n \sum (x_h - \bar{x}_h)(x_j - \bar{x}_j) = n \sum x_h x_j - \sum x_h \sum x_j$$

Note that

$$a_{hj} = a_{jh} \qquad\qquad (\text{e.g., } a_{23} = a_{32} = n \sum x_2 x_3 - \sum x_2 \sum x_3)$$

and

$$a_{jj} = n \sum x_j^2 - \left(\sum x_j\right)^2 \qquad\qquad (\text{e.g., } a_{11} = n \sum x_1^2 - \left(\sum x_1\right)^2)$$

Also let

$$a_{jy} = n \sum x_j y - \sum x_j \sum y \qquad\qquad (\text{e.g., } a_{2y} = n \sum x_2 y - \sum x_2 \sum y)$$

and

$$a_{yy} = n \sum y^2 - \left(\sum y\right)^2$$

With this notation the correlation coefficient r_{hj} between x_h and x_j can be written

$$r_{hj} = \frac{a_{hj}}{\sqrt{a_{hh} a_{jj}}}$$

For ease in computing, we consider the normal equations multiplied by n and written in the form

$$a_{11}b_1 + a_{12}b_2 + \cdots + a_{1k}b_k = a_{1y}$$
$$a_{21}b_1 + a_{22}b_2 + \cdots + a_{2k}b_k = a_{2y}$$
$$\cdots \cdots \cdots \cdots \cdots \cdots \cdots$$
$$a_{k1}b_1 + a_{k2}b_2 + \cdots + a_{kk}b_k = a_{ky}$$

References 6.7 (pp. 642–49) and 6.9 give methods for solving simultaneous linear equations. Any familiar method may be used, but a large-scale problem (more than three independent variables, or many repetitions with three or fewer) may be solved most readily by high-speed machine methods. The technique most readily adapted to machine methods and to assessing the precision of the results is that of inverting the matrix of coefficients a_{hj}. This involves solving for the b's by replacing the column of constants a_{jy} successively with 1, 0, 0, $\cdots$, 0, getting values e_{11}, e_{12}, $\cdots$, e_{1k}; with 0, 1, 0, $\cdots$, 0, getting $e_{21}, e_{22}, \cdots, e_{2k}$; $\cdots$; and with 0, 0, $\cdots$, 0, 1, getting $e_{k1}, e_{k2}, \cdots, e_{kk}$. For example, e_{11}, $e_{12}, \cdots, e_{1k}$ satisfy the equations

$$a_{11}e_{11} + a_{12}e_{12} + \cdots + a_{1k}e_{1k} = 1$$
$$a_{21}e_{11} + a_{22}e_{12} + \cdots + a_{2k}e_{1k} = 0$$
$$a_{31}e_{11} + a_{32}e_{12} + \cdots + a_{3k}e_{1k} = 0$$
$$\cdots \cdots \cdots \cdots \cdots \cdots \cdots$$
$$a_{k1}e_{11} + a_{k2}e_{12} + \cdots + a_{kk}e_{1k} = 0$$

Then the solution of the original equations is

$$b_j = \sum_{h=1}^{k} a_{hy}e_{hj} \qquad\qquad (j = 1, 2, \cdots, k)$$

Since the a_{hj} are the same in all cases, an efficient computational form, such as the one shown in Ref. 6.7 (pp. 648–49), enables us to obtain all the e_{hj} and the b_j simultaneously, and provides easy and valuable step-by-step checks.

Example. Rocket propellant grains are produced by extrusion through a die under pressure. The major inside diameter y depends not only on the die shape, but also on the kind of powder, its initial temperature x_1, the die temperature x_2, and the extrusion rate x_3. Experiments have indicated that the average influence of x_1, x_2, and x_3 on y can be adequately represented by a linear equation

$$y' = a + b_1 x_1 + b_2 x_2 + b_3 x_3$$

Table 6.2 lists 19 observations taken on a particular type of grain for the purpose of evaluating a, b_1, b_2, and b_3. The values of the x_j, except for small variations, are 110 and 130°F for x_1; 120, 140, and 158°F for x_2; and 10, 20, and 28 in/min for x_3. Thus, the problem is better classified as Type I than as Type II.

TABLE 6.2. DATA FOR MULTIPLE-REGRESSION EXAMPLE

Powder temp. − 100, °F x_1	Die temp. − 100, °F x_2	Extrusion rate in/min x_3	$[\text{ID(in.)} - 1.1]10^3$ y	$y + \sum x_j$
11	58	11	126	206
32	20	13	92	157
14	22	28	108	172
26	55	28	119	228
9	41	21	103	174
30	18	20	83	151
12	56	20	113	201
29	40	26	109	204
7	38	9	130	184
28	57	10	106	201
10	19	19	104	152
31	37	18	92	178
12	21	10	94	137
33	40	11	95	179
9	42	27	109	187
12	57	29	103	201
10	21	12	82	125
33	40	19	85	177
30	58	29	104	221

The computation is simplified by subtracting 100°F from the powder temperature and the die temperature, and by subtracting 1.100 inches from the inside diameter and expressing this last difference in thousandths of an inch. The column of sums $y + \sum x_j$ is included to provide an over-all check of the calculation of the normal equations, as shown in Table 6.3. The check is given by the identity

$$n \sum_{1}^{n} (y + \sum_{j} x_j)^2 - [\sum_{1}^{n} (y + \sum_{j} x_j)]^2$$

$$= a_{yy} + \sum_{j} a_{jj} + 2 \sum_{j} a_{jy} + 2 \sum_{h<j} \sum a_{hj}$$

In this example both sides of the identity are found by separate calculation to be 261,348.

TABLE 6.3. COMPUTATION OF SUMS OF SQUARES AND PRODUCTS

$n = 19.$

Quantity	x_1	x_2	x_3	y	$y + \sum x_j$
$\sum x_j$	378	740	360	1,957	3,435
$\sum x_j^2$	9,424	32,896	7,778	204,845	634,767
a_{jj}	36,172	77,424	18,182	62,206	261,348
$\sum x_1 x_j$		14,860	7,222	37,865	
a_{1j}		2,620	1,138	$-20,311$	
$\sum x_2 x_j$			14,505	78,213	
a_{2j}			9,195	37,867	
$\sum x_3 y$				37,247	
a_{3y}				3,173	{Check sum: 261,348
$\bar{x}_j$	19.89	38.95	18.95	103.00	

From Table 6.3, the normal equations for this example are

$$36,172b_1 + 2,620b_2 + 1,138b_3 = -20,311$$
$$2,620b_1 + 77,424b_2 + 9,195b_3 = 37,867$$
$$1,138b_1 + 9,195b_2 + 18,182b_3 = 3,173$$

By first (or simultaneously) solving the like equations in the e_{hj}'s by an appropriate method, we obtain the results

$e_{11} = 27.7444 \times 10^{-6}$　　$e_{12} = -0.7794 \times 10^{-6}$　　$e_{13} = -1.3423 \times 10^{-6}$
$e_{21} = -0.7794 \times 10^{-6}$　　$e_{22} = 13.7631 \times 10^{-6}$　　$e_{23} = -6.9115 \times 10^{-6}$
$e_{31} = -1.3423 \times 10^{-6}$　　$e_{32} = -6.9115 \times 10^{-6}$　　$e_{33} = 58.5787 \times 10^{-6}$
$b_1 = -0.597290$　　　　$b_2 = 0.515068$　　　　　$b_3 = -0.048583$

We can then write the regression equation in the form of the last equation of Sec. 6.2.1, using the mean values at the bottom of Table 6.3

$$y' - 103.00 = -0.5973(x_1 - 19.89) + 0.5151(x_2 - 38.95) \\ - 0.0486(x_3 - 18.95)$$

Multiplying and collecting terms, we can write

$$y' = 95.74 - 0.5973x_1 + 0.5151x_2 - 0.0486x_3$$

The value of the original dependent variable, the major inside diameter in inches, is predicted for given values of x_1, x_2, and x_3 by dividing the y' calculated from the above equation by 1,000 and adding 1.1. If the equation were to be used frequently, a nomogram could be constructed to give the results in chart form (Ref. 6.1). If $x_1 = 11$, $x_2 = 58$, and $x_3 = 11$, then $y' = 118.51$ and the predicted major inside diameter is 1.2185 inches. This predicted value may be compared with the first observed value, 1.226 inches (coded as 126) in Table 6.2. The regression equation gives a prediction of the *average* value of y for the given values of x_1, x_2, and x_3, as shown in Fig. 6.1 and 6.4. Individual observations, future as well as past if background conditions remain unchanged, will tend to cluster about the sample regression plane if sufficient data are used to make it a good approximation to the population regression plane. The usefulness of the equation is examined in the following sections.

6.2.3. Variation About the Regression Plane

The concepts associated with multiple regression analysis are to a great extent similar to those associated with a simple regression line. In particular, the scatter in the vertical (y) direction of the observed points about the regression plane is measured by the **standard error of estimate** $s_{y|12\cdots k}$, where

$$s^2_{y|12\cdots k} = \frac{1}{n-k-1}\sum_{i=1}^{n}(y_i - y'_i)^2$$

It can be calculated as

$$s^2_{y|12\cdots k} = \frac{1}{n(n-k-1)}(a_{yy} - b_1 a_{1y} - b_2 a_{2y} - \cdots - b_k a_{ky})$$

or, as will be seen in Sec. 6.2.4, as

$$s^2_{y|12\cdots k} = \frac{(n-1)s^2_y}{n-k-1}(1 - r^2_{y|12\cdots k})$$

where $r_{y|12\ldots k}$ is the multiple correlation coefficient.

If $s_{y|12\ldots k}$ is based on a sufficiently large sample, it is a good estimate of the scatter of the population about the true or population regression plane. If the deviations from the plane are normally distributed, about 95% of the points in a large sample will lie within $\pm 2s_{y|12\ldots k}$ of the plane (measured in the y direction).

Example. For the example of Table 6.2, we calculate

$$s_{y|123}^2 = \frac{1}{(19)(15)} [62,206 - (-0.597290)(-20,311)$$

$$- (0.515068)(37,867) - (-0.048583)(3,173)]$$

$$= 107.81$$

$$s_{y|123} = 10.383 \times 10^{-3} \text{ inch}$$

whereas

$$s_y^2 = \frac{a_{yy}}{n(n-1)} = \frac{62,206}{(19)(18)} = 181.89$$

$$s_y = 13.487 \times 10^{-3} \text{ inch}$$

If the deviations are approximately normally distributed, about 68%, or 13, of the 19 observed inside diameters should lie within 0.0104 inch of the values predicted from the corresponding x_j values by the regression equation; actually 12 do so. Likewise about 95%, or 18, should lie within 0.0208 inch, and 19 do so.

6.2.4. Relative Variation (Type II Problem)

a. Multiple Correlation Coefficient. Just as in the case of one independent variable, the **population multiple correlation coefficient** squared, $\rho_{y|12\ldots k}^2$, is defined as the fraction of the total variance of y that is accounted for by its regression on the variables $x_1, x_2, \cdots$, and x_k. The best estimate of $\rho_{y|12\ldots k}^2$ from the sample is the **sample multiple correlation coefficient** squared

$$r_{y|12\ldots k}^2 = \frac{b_1 a_{1y} + b_2 a_{2y} + \cdots + b_k a_{ky}}{a_{yy}}$$

The multiple correlation coefficient itself is always taken as the positive square root of $r_{y|12\ldots k}^2$, and it is never greater than one: $0 \leqslant r_{y|12\ldots k} \leqslant 1$.

A value of zero indicates no correlation between y and x_j; a value of one means that all sample points lie precisely on the regression plane. Since, because of random fluctuations, $r_{y|12\ldots k}$ is almost never zero even when there is no correlation between y and the x_j in the population, it is important to test its statistical significance; this is easily done as shown in Sec. 6.2.5b.

As in the case of simple correlation in Sec. 6.1.3, the multiple correlation coefficient is appropriate only in Type II problems (in which the variables x_j vary randomly rather than as chosen by the experimenter). In both Type I and Type II problems, the size of $s_{y|12\ldots k}$ is used to assess the goodness of fit (Sec. 6.2.3).

No numerical example will be given here since the numerical data in Table 6.2 are for a Type I problem.

b. Partial Correlation Coefficients. This subsection, like Subsec. a, is appropriate only for Type II problems. The **sample partial correlation coefficient** $r_{y3|124\ldots k}$, for example, estimates the correlation between y and x_3 after the influences of x_1, x_2, x_4, $\cdots$, and x_k, as estimated from the sample, have been removed. The partial correlation coefficient squared, $r^2_{y3|124\ldots k}$, is approximately the fraction of the otherwise unexplained part of the variance of y that will be accounted for by considering the remaining variable x_3. The meaning of the partial correlation coefficient should be sharply distinguished from that of the simple correlation coefficient r_{y3}; the former systematically eliminates, but the latter merely ignores, the variation caused by x_1, x_2, x_4, $\cdots$, and x_k.

For a three-variable regression ($k = 2$) we can calculate the partial correlation coefficients from the simple correlation coefficients r_{hj} by the formula

$$r_{y2|1} = \frac{r_{y2} - r_{y1}r_{21}}{\sqrt{(1 - r^2_{y1})(1 - r^2_{21})}}$$

merely interchanging the 1 and the 2 to obtain $r_{y1|2}$. (See Sec. 6.2.2 for calculation of the r_{hj}.) This formula shows that r_{y2} may be positive while $r_{y2|1}$ is zero or negative, so that little or nothing about *basic* relations can be concluded directly from the simple correlation coefficient if there is some further variable influencing y. For example, if $r_{y2} = 0.5$, and $r_{y1} = r_{21} = 0.8$, then $r_{y2|1} = -0.389$.

For a four-variable regression we have

$$r_{y3|12} = \frac{r_{y3|1} - r_{y2|1}\, r_{32|1}}{\sqrt{\left(1 - r^2_{y2|1}\right)\left(1 - r^2_{32|1}\right)}}$$

and similarly for the other coefficients. In this formula, the three coefficients $r_{hj|1}$ on the right must be calculated from the six simple coefficients r_{hj}.

Each partial correlation coefficient lies between -1 and $+1$. The interpretation of the coefficient is analogous to that of the correlation coefficient for two variables. Significance tests and confidence intervals can also be constructed as in the case of two variables (Sec. 6.2.5c and d). In most applications it is sufficient to calculate the partial *regression* coefficients, the b_j, and not bother with the partial *correlation* coefficients.

6.2.5. Reliability of Multiple Regression Measures

As in simple regression (Sec. 6.1.4), the significance tests and confidence intervals given below are exact only under the normality assumptions discussed in Sec. 6.0. Subsections a, e, f, and g are appropriate to Type I problems. Numerical examples are not given for the other subsections, since the problem of Table 6.2 is better classified as Type I than as Type II. All of this section may be applied to Type II problems, but Subsec. b and c are equivalent to special cases of Subsec. a and e, respectively. Problems of mixed type may arise in multiple regression.

a. Significance of Regression as a Whole. The regression plane could fail to be statistically significant because (1) the assumed form of the equation is not the true form (i.e., the true regression surface is not a plane), or (2) the variation of observed points about the fitted plane, though random, is so large that the fitted plane could have arisen by random sampling from a population with all the partial regression coefficients equal to zero.

We can apply the method of Sec. 6.1.6 to test the *linearity* of the regression relation if several observations of y are made for each combination of values of $x_1, x_2, \cdots, x_k$.

We can test the hypothesis that all true partial regression coefficients equal zero by an F test of the variance accounted for by regression, relative to the error variance $s^2_{y|12\cdots k}$

$$F = \frac{[b_1 \sum (x_{1i} - \bar{x}_1)(y_i - \bar{y}) + \cdots + b_k \sum (x_{ki} - \bar{x}_k)(y_i - \bar{y})]/k}{s^2_{y|12\cdots k}}$$

$$= \frac{b_1 a_{1y} + \cdots + b_k a_{ky}}{kns^2_{y|12\cdots k}}$$

with k and $n - k - 1$ degrees of freedom. We compare the F calculated from the data with the critical value tabulated in Appendix Table 5, and reject the hypothesis that all true partial regression coefficients equal zero if the calculated F is larger.

Example. Test at the 5% significance level whether the true partial regression coefficients are zero in the example of Table 6.2.

We find from Sec. 6.2.3 that

$$F = \frac{31,481}{(3)(19)(107.81)} = 5.123$$

with 3 and 15 degrees of freedom, which is larger than the critical value $F_{.05}(3, 15) = 3.29$ in Appendix Table 5. Hence we reject the hypothesis, concluding that the variance accounted for by regression is more than could reasonably be expected if all the true partial regression coefficients were zero.

b. Significance of Multiple Correlation Coefficient. We can test $r_{y|12\cdots k}$ for significance at the 1 or 5% level by comparison with the value in Appendix Table 7 for $k + 1$ variables and $n - k - 1$ degrees of freedom. We reject the null hypothesis that the population multiple correlation coefficient is zero if $r_{y|12\cdots k}$ exceeds the tabled value. If we reject the null hypothesis, we conclude that the regression of y on $x_1, x_2, \cdots,$ and x_k accounts for a significant amount of the variation observed in y. This test of significance of the multiple correlation coefficient is actually equivalent to the F test of Sec. 6.2.5a.

c. Significance of Partial Correlation Coefficients. We can test any partial correlation coefficient, $r_{y1|23\cdots k}$ for instance, for significance at the 1 or 5% level by comparison with the value in Appendix Table 7 for $n - k - 1$ degrees of freedom and *two* variables. If $|r_{y1|23\cdots k}|$ exceeds the tabled value, we reject the null hypothesis that the population partial correlation coefficient $\rho_{y1|23\cdots k}$ is zero. Note that, although we use the table column for two variables (the variables x_1 and y, in this case), we take care of the deletion of the other variables by using $n - k - 1$ degrees of freedom. Thus, for $k = 3$ and $n = 19$ the critical value of $|r_{y1|23}|$ at the 5% significance level is 0.482.

d. Confidence Intervals for Partial Correlation Coefficients. This subsection, like the preceding two, applies only to Type II problems. To construct a 95% confidence interval for the population partial correlation coefficient, say $\rho_{y3|124\ldots k}$, enter Appendix Chart XI with the sample coefficient $r_{y3|124\ldots k}$ and find the interval cut off by the curves labeled $n - k + 1$. In 95% of the experiments so treated, the resulting confidence interval will contain the population coefficient. Thus, if $k = 3$, $n = 19$, and $r_{y3|12} = 0.6$, we interpolate for 17 between curves labeled 15 and 20 and get 95% confidence limits of 0.16 and 0.83.

e. Significance Tests and Confidence Intervals for Partial Regression Coefficients. The standard error of the partial *regression* coefficient b_j can be estimated from the sample as

$$s_{b_j} = s_{y|12\ldots k}\sqrt{ne_{jj}} \qquad (j = 1, 2, \cdots, k)$$

The values of the e_{jj} are the diagonal elements of the inverse matrix of the normal equations computed in Sec. 6.2.2.

To test the null hypothesis that the partial regression coefficient β_j in the population has any stated value B, compute

$$t = \frac{b_j - B}{s_{b_j}}$$

Reject the null hypothesis at the significance level α if $|t|$ exceeds the critical value $t_{\alpha/2,n-k-1}$ given in Appendix Table 3 for $n - k - 1$ degrees of freedom.

A $100(1 - \alpha)\%$ *confidence interval* for the population coefficient β_j is

$$b_j \pm t_{\alpha/2,n-k-1}s_{b_j}$$

Example. In the continuing example, we have

$b_1 = -0.597290$ $e_{11} = 27.7444 \times 10^{-6}$ $s_{y|123} = 10.383$

$b_2 = 0.515068$ $e_{22} = 13.7631 \times 10^{-6}$ $n = 19$

$b_3 = -0.048583$ $e_{33} = 58.5787 \times 10^{-6}$

Hence

$$s_{b_1} = 10.383 \sqrt{5.2714 \times 10^{-4}} = 0.23839$$

$$s_{b_2} = 10.383 \sqrt{2.6150 \times 10^{-4}} = 0.16790$$

$$s_{b_3} = 10.383 \sqrt{11.1300 \times 10^{-4}} = 0.34639$$

To test at the 5% significance level the null hypothesis that a population partial regression coefficient β_j is zero, we compute $t_j = b_j/s_{b_j}$ and compare its absolute value with $t_{.025,15} = 2.131$ from Appendix Table 3. Since b_3 is much smaller numerically than its estimated standard error, it is not significant (i.e., not significantly different from zero) whatever reasonable significance level might have been chosen, but

$$t_1 = -2.506 \qquad \text{and} \qquad t_2 = 3.068$$

Hence we reject the hypotheses that $\beta_1 = 0$ and $\beta_2 = 0$.

We may construct 95% confidence limits for β_1 as

$$b_1 \pm 2.131 s_{b_1} = -0.5973 \pm 0.5080$$

$$= -1.1053 \text{ and } -0.0893$$

Likewise 95% confidence limits for β_2 are

$$b_2 \pm 2.131 s_{b_2} = 0.5151 \pm 0.3578$$

$$= 0.1573 \text{ and } 0.8729$$

Hence β_1 and β_2 are but roughly determined by the 19 observations available.

In general, when a partial regression coefficient, say b_h, is not statistically significant, we may choose one of three courses of action:

1. Retain the term $b_h(x_h - \bar{x}_h)$ in the regression equation

2. Calculate a new regression equation, preferably on the basis of a further sample, omitting observations on x_h entirely

3. Discard the term and use the regression equation with the other terms already calculated

The first course is the safest, but it may be unnecessarily complicating when, as in the example below, the term can be discarded on the basis of its negligible practical importance. If a term is discarded, course 2 is the proper one in general, in order that the resulting reduced equation be a least-squares estimate of the (reduced) population regression plane. If the experiment is designed with such symmetry that the off-diagonal coefficients a_{hj} $(h \neq j)$ of the normal equations are zero, following course 3 is equivalent to following course 2 without drawing a new sample. Also, course 3 is sometimes justified as an approximation to course 1 when, as in the example, the differences between the y''s as predicted by the resulting reduced equation and by the equation using all the variables are negligible.

In any case the choice of equation is aided by comparison of the sizes of the standard errors of estimate $s_{y|12\ldots k}$ of the various equations considered. In an extreme case the regression as a whole may be significant according to Sec. 6.2.5a, although no b_j is significant individually. It may happen in such a case that some subset of the original set of independent variables yields practically as useful a regression as the original. This is exemplified in Ref. 6.7, p. 646, together with a method of finding the appropriate subset of independent variables.

Example. In the previous example, we found that two regression coefficients, b_1 and b_2, were significant, but the third, b_3, was not significant. What would the practical effect in estimating y be if we were to discard the term in x_3?

Over the range of applicability of the regression, the maximum error introduced by discarding the term $b_3 (x_3 - \bar{x}_3)$ would be $(0.0486)(10.05)(10^{-3}) = 0.00049$ inch. Therefore, we should follow course 3 of the explanation above, discarding the term because its contribution is not only not significant statistically but is also negligible numerically.

f. Confidence Interval for Ordinate to the True Regression Plane.
Confidence limits for the true *mean* y value (*not* for an individual predicted y) at the x point $(X_1, X_2, \cdots, X_k)$ are

$$y' \pm t_{a/2,n-k-1} s_{y|12\ldots k} \sqrt{\frac{1}{n} + n \sum_{h,j=1}^{k} e_{hj} (X_h - \bar{x}_h)(X_j - \bar{x}_j)}$$

where y' is calculated from the regression equation. The x point need not be one of those used in finding the regression plane. Note that there are k^2 terms in the summation, k squares plus two equal cross-products for each pair (h, j).

This confidence interval can be particularized to cover the case of the y intercept by taking $(X_1, X_2, \cdots, X_k) = (0, 0, \cdots, 0)$ or the case of $\bar{y}$ by taking $(X_1, X_2, \cdots, X_k) = (\bar{x}_1, \bar{x}_2, \cdots, \bar{x}_k)$. Significance tests based on the t statistic can also be made.

Example. For the data of Table 6.2, calculate 95% confidence limits for the mean y value ($y' = 118.51$ or $(ID)' = 1.2185$ inches) estimated in Sec. 6.2.2 for $x_1 = 11$, $x_2 = 58$, and $x_3 = 11$.

Using the values of $\bar{x}_j$ in Table 6.3 and the e_{hj} calculated in Sec. 6.2.2, we have

$$n \sum_{h,j=1}^{3} e_{hj} (X_h - \bar{x}_h)(X_j - \bar{x}_j)$$

$$= 19 \times 10^{-6} [27.74(-8.89)^2 + 13.76(19.05)^2$$

$$+ 58.58(-7.95)^2 + 2(-0.78)(-8.89)(19.05)$$

$$+ 2(-1.34)(-8.89)(-7.95)$$

$$+ 2(-6.91)(19.05)(-7.95)]$$

$$= 0.2481$$

Hence the confidence limits are

$$118.51 \pm (2.131)(10.383)(0.5484) = 118.51 \pm 12.13$$

$$= 106.38 \text{ and } 130.64$$

or 1.2064 and 1.2306 inches, in terms of the original scale for inside diameter.

g. Prediction Interval for Individual Value of y. The appropriate discussion here is the same as in Sec. 6.1.4f. A $100(1 - \alpha)\%$ prediction interval for an individual value of y for the x point $(X_1, X_2, \cdots, X_k)$ differs from the $100(1 - \alpha)\%$ confidence interval for the ordinate to the true regression plane given in Sec. 6.2.5f only in replacing the first term under the radical, $1/n$, by $1 + 1/n$.

Example. For the data of Table 6.2, calculate a 95% prediction interval for an individual value of y for $x_1 = 11$, $x_2 = 58$, and $x_3 = 11$.

The calculation involves the summation evaluated in the example in Sec. 6.2.5f. The prediction limits are

$$118.51 \pm (2.131)(10.383)(1.1405) = 118.51 \pm 25.23$$
$$= \ \ 93.28 \text{ and } 143.74$$

or 1.1933 and 1.2437 inches, in terms of the original scale.

6.3. Non-Linear Regression

This section applies to Type I problems but not to Type II problems (Sec. 6.0). As noted in Sec. 6.0, the method generally used for fitting a line, curve, or surface is the *method of least squares*. If, except for random fluctuation in y (which need not be normally distributed), y is related to variables $x_1, x_2 \cdots , x_k$ by a function $g(x_1, x_2, \cdots , x_k; c_1, c_2, \cdots , c_m)$ of *given* form with unknown parameters $c_1, c_2, \cdots , c_m$, then these parameters may be estimated from observations $x_{1i}, x_{2i}, \cdots , x_{ki}, y_i$ (where $i = 1, 2, \cdots , n$) which have known "weights" w_i by minimizing the weighted sum of squared deviations

$$S(c_1, c_2, \cdots , c_m) = \sum_{i=1}^{n} [y_i - g(x_{1i}, x_{2i}, \cdots , x_{ki}; \ c_1, c_2, \cdots , c_m)]^2 w_i$$

with respect to $c_1, c_2, \cdots , c_m$. (Throughout this Manual all $w_i = 1$.) This minimization is generally done by differentiating $S(c_1, c_2, \cdots , c_m)$ with respect to each c_j and equating to zero, thus getting m equations for estimating $c_1, c_2, \cdots , c_m$.

Many non-linear relationships can be analyzed by transforming the original variables into new ones that are related linearly. Thus, to fit a cubic equation

$$y' = a + b_1 x + b_2 x^2 + b_3 x^3$$

we let $x_1 = x$, $x_2 = x^2$, and $x_3 = x^3$. With these meanings for x_1, x_2, and x_3, the formulas for multiple regression in Sec. 6.2 may be used immediately for non-linear regression. The intercept on the y axis is estimated as

$$a = \bar{y} - b_1 \bar{x} - b_2 \overline{x^2} - b_3 \overline{x^3}$$

where $\overline{x^2}$ is the mean of the squares of the observed values of x and $\overline{x^3}$ is the mean of the cubes of the observed values. We note that x_1, x_2, and x_3 are certainly not statistically independent here, but independence is not necessary in multiple regression analysis.

The methods detailed in Sec. 6.1 and 6.2 can be used whenever $y' = g(x_1, x_2, \cdots, x_k; c_1, c_2, \cdots, c_m)$ is linear in the parameters to be estimated. For example, the equation

$$y' = a + b_1\sqrt{z_1} + b_2 z_1 \log z_2 + b_3/z_2^2$$

can be fitted, using the formulas of Sec. 6.2, by letting $x_1 = \sqrt{z_1}$, $x_2 = z_1 \log z_2$, and $x_3 = 1/z_2^2$. The y intercept is estimated as

$$a = \bar{y} - b_1\overline{\sqrt{z_1}} - b_2 \overline{z_1 \log z_2} - b_3 \overline{(1/z_2^2)}$$

In particular, the test of a hypothetical form of an equation in Sec. 6.1.6 is available if, as is recommended, several observations on y can be made for each value of $x_1, x_2, \cdots, x_k$, and if the random fluctuations in y are normally distributed.

Sometimes a transformation of y' as well as x (or the x_j in multiple regression) may render y' linear in the parameters. For example, $y'x^b = c$ becomes

$$\log y' = \log c - b \log x$$

by taking logarithms. By letting $Y' = \log y'$, $C = \log c$, and $X = \log x$, the problem is reduced to the linear case of Sec. 6.1. Other transformations of value are

$$X = e^{ax} \qquad \text{and} \qquad X = x^a$$

The latter transformation includes $X = 1/x^b$ and $X = \sqrt[c]{x}$, since a can be any known non-zero constant. Similar changes of variable may be made on y'.

If a transformation of y' is used, say to $\log y'$, the fitted line or plane minimizes the sum of squares of deviations of $\log y$ rather than of y; if y is normally distributed, $\log y$ will not be. Although in this case the theoretically best estimate of the regression curve of y on the x's is not obtained, the approximation is often sufficiently close. In case the regression function cannot be transformed so that the unknown coefficients are involved linearly, the least-squares equations may be insoluble except by a method of successive approximations. (See Ref. 6.12 and also Ref. 6.7, pp. 558–70 and 649–57, for further discussion of the fitting of non-linear equations.)

6.3.1. Polynomial Equations

For curvilinear relationships, a polynomial is frequently fitted unless there is some theoretical basis for using a different curve. The polynomial equation is

$$y' - \bar{y} = b_1 (x - \bar{x}) + b_2 (x^2 - \overline{x^2}) + \cdots + b_k (x^k - \overline{x^k})$$

where

$$\overline{x^j} = \frac{1}{n} \sum_{i=1}^{n} x_i^j \qquad\qquad (j = 1, 2, \cdots, k)$$

The normal equations of Sec. 6.2.2 for determining the coefficients hold as before

$$a_{11}b_1 + a_{12}b_2 + \cdots + a_{1k}b_k = a_{1y}$$
$$a_{21}b_1 + a_{22}b_2 + \cdots + a_{2k}b_k = a_{2y}$$
$$\cdot \quad \cdot \quad \cdot \quad \cdot \quad \cdot \quad \cdot \quad \cdot \quad \cdot \quad \cdot \quad \cdot \quad \cdot \quad \cdot$$
$$a_{k1}b_1 + a_{k2}b_2 + \cdots + a_{kk}b_k = a_{ky}$$

where now

$$a_{hj} = n \sum x^{h+j} - \sum x^h \sum x^j$$
$$a_{jy} = n \sum x^j y - \sum x^j \sum y$$

As in Sec. 6.2.2, the equations may be solved for the b_j in the easiest available way. All the methods of Sec. 6.2 for a Type I model can be applied directly to the polynomial equation and need not be repeated here. In particular, the variance about the regression curve is defined as

$$s_{y|x}^2 = \frac{1}{n - k - 1} \sum_{i=1}^{n} (y_i - y_i')^2$$

and is calculated as

$$s_{y|x}^2 = \frac{1}{n(n - k - 1)} (a_{yy} - b_1 a_{1y} - b_2 a_{2y} - \cdots - b_k a_{ky})$$

where

$$a_{yy} = n \sum y^2 - (\sum y)^2$$

Since the present situation is simply that of multiple linear regression with x_j replaced by x^j, the significance of any regression coefficient can be tested by the method of Sec. 6.2.5e. If there is uncertainty about the degree of equation appropriate for the data, it is desirable to start with an equation of the highest possible degree of interest, and subsequently to discard those terms which contribute little to the total estimate y' and have coefficients that are not significant. In this situation the use of orthogonal polynomials (described in the next section) is convenient. The method does not require further calculation for the reduced equation; however, its application is practical only if the values of x are equally spaced.

6.3.2. Orthogonal Polynomials, When x Is Equally Spaced

When an ordinary polynomial is to be fitted (Sec. 6.3.1), all the coefficients must be recalculated each time a power term is added or discarded. With **orthogonal polynomials** an analysis of variance is performed to apportion the variance among terms of the polynomial. The contribution of each term is tested; if it is not found significant, the term may be discarded without recalculating the previously obtained coefficients. For fitting curves of higher degree, this method saves time. Furthermore, the tests of significance of effects are isolated, as in the simpler cases of analysis of variance.

The orthogonal polynomials themselves depend on the arrangement of the values of the independent variable x. For the commonly tabulated orthogonal polynomials used here, the values of x must be equally spaced. The method yields an equation

$$y' = \bar{y} + A_1\xi_1'(x) + A_2\xi_2'(x) + \cdots + A_k\xi_k'(x)$$

where the A_j are coefficients to be determined, and the $\xi_j'(x)$ are known polynomials in x—orthogonal in the sense that the products $\xi_h'\xi_j'$ (with $h \neq j$) over n equally spaced values of x sum to zero.

In order to tabulate the values of the orthogonal polynomials conveniently for repeated use, the values of x—say x_i—are taken to be one unit apart, and $\xi_j'(x)$ is taken as a multiple λ_j of a corresponding polynomial $\xi_j(x)$ with leading coefficient unity. The latter adjustment makes all the tabulated values $\xi_j'(x_i)$ integers. Thus, in particular, we have

$$\xi_1'(x) = \lambda_1 \xi_1(x) = \lambda_1 (x - \bar{x})$$

$$\xi_2'(x) = \lambda_2 \xi_2(x) = \lambda_2 \left[(x - \bar{x})^2 - \frac{n^2 - 1}{12} \right]$$

$$\xi_3'(x) = \lambda_3 \xi_3(x) = \lambda_3 \left[(x - \bar{x})^3 - (x - \bar{x}) \frac{3n^2 - 7}{20} \right]$$

$$\xi_4'(x) = \lambda_4 \xi_4(x) = \lambda_4 \left[(x - \bar{x})^4 - (x - \bar{x})^2 \frac{3n^2 - 13}{14} + \frac{3(n^2 - 1)(n^2 - 9)}{560} \right]$$

and in general $\xi_r'(x) = \lambda_r \xi_r$ with the recursion formula

$$\xi_{r+1} = \xi_1 \xi_r - \frac{r^2(n^2 - r^2)}{4(4r^2 - 1)} \xi_{r-1}$$

where

$$\xi_0 = 1 \qquad \text{and} \qquad \xi_1 = x - \bar{x}$$

The coefficients A_j of a fitted polynomial are calculated from the observations y_i, using the formula

$$A_j = \frac{\sum y_i \xi_j'(x_i)}{\sum \xi_j'^2(x_i)}$$

The values of the individual $\xi_j'(x_i)$ and the sums of squares $\sum \xi_j'^2(x_i)$ are given in tables of orthogonal polynomials. Appendix Table 15, which is an excerpt from a more extensive table given in Ref. 5.7, illustrates the arrangement of that table and suffices for the example below; the complete table of Ref. 5.7 would have to be consulted in general. The tables also give values of the λ_j, which depend on n; with these y' can be expressed directly in powers of x.

Example. From the data in columns 2 and 3 of Table 6.4, fit a polynomial curve. The values of the independent variable x are coded one unit apart in column 2 of the table. It is assumed that terms of degree higher than four are known from theory not to be present or are certain to be of negligible interest.

Columns 4–7 and the values of $\sum \xi_j'^2$ and λ_j are taken from the table of orthogonal polynomials, Appendix Table 15, for $n = 15$. The values for $x \geqslant 8$ are given directly in that table. The values for $x < 8$ are the same (symmetrically about $\bar{x}$) for ξ_2' and ξ_4', and the same except for

TABLE 6.4. EXAMPLE OF USE OF ORTHOGONAL POLYNOMIALS

x_i Original	x_i Coded	y_i	ξ_1'	ξ_2'	ξ_3'	ξ_4'
(1)	(2)	(3)	(4)	(5)	(6)	(7)
45.2	1	7.8	−7	91	−91	1,001
45.4	2	7.9	−6	52	−13	−429
45.6	3	8.8	−5	19	35	−869
45.8	4	9.3	−4	−8	58	−704
46.0	5	8.7	−3	−29	61	−249
46.2	6	8.4	−2	−44	49	251
46.4	7	7.8	−1	−53	27	621
46.6	8	7.2	0	−56	0	756
46.8	9	6.7	1	−53	−27	621
47.0	10	6.4	2	−44	−49	251
47.2	11	5.7	3	−29	−61	−249
47.4	12	4.6	4	−8	−58	−704
47.6	13	3.7	5	19	−35	−869
47.8	14	3.0	6	52	13	−429
48.0	15	2.0	7	91	91	1,001
$\sum \xi_j'^2$	....		280	37,128	39,780	6,466,460
λ_j...............	....		1	3	$\frac{5}{6}$	$\frac{35}{12}$
$\sum y_i \xi_j'$		98.0	−128.4	−655.6	170.3	−937.5

change of sign for ξ_1' and ξ_3'. The sums of products $\sum y_i \, \xi_j' \, (x_i)$ must be calculated and are given at the bottom of Table 6.4. The coefficients in the regression equation are then obtained immediately as

$$\bar{y} = 6.5333$$

$$A_1 = \frac{-128.4}{280} = -0.45857 \qquad A_3 = \frac{170.3}{39,780} = 0.0042810$$

$$A_2 = \frac{-655.6}{37,128} = -0.017658 \qquad A_4 = \frac{-937.5}{6,466,460} = -0.00014498$$

An analysis of variance (Chapter 5) is performed to test the significance of the terms $A_j \, \xi_j' \, (x)$. We first compute the sum of squared deviations of y from $\bar{y}$, which is the *total* sum of squares; then in turn we compute that part of the total sum of squares accounted for by the linear, quadratic, cubic, and quartic regressions. These sums of squares, their degrees of freedom, and the corresponding mean squares appear in Table 6.5. The sum of squares accounted for by the jth degree term $A_j \xi'_j(x)$ is calculated as

$$\frac{(\sum y_i \xi'_j)^2}{\sum \xi'^2_j}$$

The residual sum of squares, that part of the total not accounted for at all except as random fluctuation, is obtained by subtraction as shown in the table.

In testing the jth degree polynomial term for significance, we are testing the null hypothesis that the population has a polynomial regression with the true value of the coefficient A_j equal to zero. Rejection of this hypothesis implies that the data are significantly better represented by a polynomial which includes $\xi'_j(x)$ than by a polynomial which does not. (It is possible, though perhaps not likely, that we might reject, for example, the hypothesis that $A_4 = 0$ and not reject the hypothesis that $A_2 = 0$.) We compare the F values with $F_{.05}(1, 10) = 4.96$ from Appendix Table 5, and we see that all terms but the quartic are significant. Therefore we use only the first three in the equation

$$y' = \bar{y} + A_1 \xi'_1(x) + A_2 \xi'_2(x) + A_3 \xi'_3(x)$$

With $n = 15$, $\lambda_1 = 1$, $\lambda_2 = 3$, and $\lambda_3 = \frac{5}{6}$ from Table 6.4 we obtain

$$\xi'_1(x) = x - 8$$

$$\xi'_2(x) = 3x^2 - 48x + 136$$

$$\xi'_3(x) = \frac{1}{6}(5x^3 - 120x^2 + 793x - 1{,}224)$$

Substitution of these and the values for $\bar{y}$, A_1, A_2, and A_3 in the equation for y' yields

$$y' = 0.00357x^3 - 0.1386x^2 + 0.955x + 6.93$$

where x is the coded variable. The equation is plotted in Fig. 6.5 along with the original data.

TABLE 6.5. TESTS OF SUMS OF SQUARES ACCOUNTED FOR BY ORTHOGONAL POLYNOMIALS

Quantity	Sum of squares	Degrees of freedom	Mean square	$F = \dfrac{\text{mean square}}{\text{residual mean square}}$
(1) $\dfrac{n\sum y^2 - (\sum y)^2}{n}$	72.0333	$n - 1 = 14$		
(2) Linear regression: $\dfrac{(\sum y_i \xi_1)^2}{\sum \xi_1^2}$	58.8806	1	58.8806	827.9
(3) Quadratic regression: $\dfrac{(\sum y_i \xi_2)^2}{\sum \xi_2^2}$	11.5765	1	11.5765	162.8
(4) Cubic regression: $\dfrac{(\sum y_i \xi_3)^2}{\sum \xi_3^2}$	0.7291	1	0.7291	10.25
(5) Quartic regression: $\dfrac{(\sum y_i \xi_4)^2}{\sum \xi_4^2}$	0.1359	1	0.1359	1.911
(6) Residual sum of squares = (1) − (2) − (3) − (4) − (5)	0.7112	$14 - 1 - 1 - 1 - 1 = 10$	0.07112	

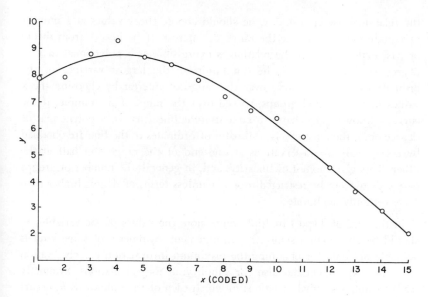

FIG. 6.5. Cubic Fitted to Data of Table 6.4.

6.4. Planning the Experiment for Fitting a Function

The general steps in designing and running an experiment outlined in Sec. 5.1, as well as much of the discussion of Sec. 5.0, apply to the fitting of functions. In particular, we should use one of the following procedures: (1) hold background variables constant at the values of interest (thus limiting the applicability of the results), (2) provide for including background conditions as independent variables in the analysis by controlling or measuring them, or (3) randomize the effects of background conditions on the observed variables. For instance, in estimating the regression function of y on one independent variable x, it would be poor technique to begin with the smallest value of x and proceed to successively larger values, because other variables affecting y (such as temperature or instrument calibration) may change with time in a systematic manner. It would be better to randomize the order of application of the values of x by consulting a random-number table (Ref. 1.7).

In a Type I problem the experimenter chooses the values of x (or of $x_1, x_2, \cdots, x_k$) for which to observe y. If he does not know the form of the relation between y and x, he should choose the x values at a number of equidistant points over the range of interest. If he knows from theory or past experience that the relation is expressible as a polynomial in x of degree no higher than k, he can minimize the largest variance of the ordinate to the fitted curve over the range of interest by choosing the x values in $k + 1$ equal groups, spaced over the range. For example, if the curve is known beforehand to be a straight line (that is, a polynomial of degree one), the most precise estimates of ordinates to the line are obtained by taking half the observations at one end of the range and half at the other. This gives no test of linearity; and, in general, the number of groups of x's should not be restricted to $k + 1$ unless terms of degree higher than k are certainly negligible.

In the case of Type I multiple regression, the values of the variables x_j should be chosen in a symmetric arrangement. Symmetry of some kind is valuable not only for obtaining the maximum information per observation from the experiment but also for simplifying the calculations. Symmetry leads to simple normal equations in which each of the unknowns b_j occurs in only one of the normal equations so that it can be solved for immediately. Furthermore, the t test of significance of each b_j will be independent of effects of the other x_j variables; this desirable result is similar to that found in fitting by use of orthogonal polynomials (Sec. 6.3.2), and in the analysis of variance of an experiment designed with appropriate symmetry (Chapter 5).

In the case of Type II problems it may happen that a large number of independent variables are present and observable, so that the question may arise as to whether they should all be observed and included in the analysis. Inclusion of a further independent variable in the analysis will always decrease the sum of squared deviations of the data from the fitted regression function. However, if the additional variable actually has little effect on the dependent variable, its inclusion may not decrease the mean square deviation, since the number of degrees of freedom is decreased also. The result is that the experiment is less sensitive; that is, all F ratios are decreased and their critical values are increased.

Moreover, if regression analysis is being applied to a Type II problem, two variables x_j may be highly correlated; if so, one of them should be omitted from the regression equation because the two regression coefficients would be poorly determined, whereas one by itself can be well determined. An extreme example might occur in a regression on percentage composition of some measurable characteristic of a solid propellant. Each of the components might conceivably be taken as an x_j, but if the percentage of one were determined as a difference between 100 and the sum of the other percentages, it would be linearly dependent on the others, the determinant of the normal equations would be zero, and the regression coefficients could not be determined. One of the components should therefore be omitted from the regression equation.

It is important to take a sufficiently large sample to determine the regression function with the desired precision, but it is somewhat complicated to predict the sample size. In addition to the form of the regression equation, one must know approximately the population standard deviation $\sigma_{y|x}$ about the regression function and the spacing of x values to be used. For example, with a straight line and the values of x equally distributed at the two end points of the range of x, the first equation of Sec. 6.1.4c indicates that the standard deviation of the estimate of the regression coefficient is

$$\sigma_b = \frac{\sigma_{y|x}}{L\sqrt{n}}$$

where $2L$ is the total range of x. Knowing $\sigma_{y|x}$, L, and the desired precision σ_b in estimating the slope, we can solve this equation for the number n of observations.

The design of experiments for regression is treated in Ref. 6.7, especially on pages 536 and 632–34.

BIBLIOGRAPHY

6.1. Arkin, H., and R. R. Colton. Graphs, How To Make and Use Them. New York, Harper, 1940. Chap. XII. Construction of nomograms.

6.2. Cramér, H. Mathematical Methods of Statistics. Princeton, N. J., Princeton University Press, 1946. Pp. 301–8, 394–415, 548–56.

6.3. Croxton, F. E., and D. J. Cowden. Applied General Statistics. New York, Prentice-Hall, 1946. Pp. 651–789.

6.4. Dixon, W. J., and F. J. Massey, Jr. An Introduction to Statistical Analysis. New York, McGraw-Hill, 1951. Chap. 11–12.

6.5. Ezekiel, M. Methods of Correlation Analysis, 2nd ed. New York, Wiley, 1941.

6.6. Frankford Arsenal. Statistical Manual. Methods of Making Experimental Inferences, 2nd ed., by C. W. Churchman. Philadelphia, Frankford Arsenal, June 1951. Pp. 88–92. Orthogonal polynomials.

6.7. Hald, A. Statistical Theory With Engineering Applications. New York, Wiley, 1952. Chap. 18–20.

6.8. Kenney, J. F. Mathematics of Statistics, Part One, 2nd ed. New York, Van Nostrand, 1947. Pp. 170–227.

6.9. Milne, W. E. Numerical Calculus. Princeton, N. J., Princeton University Press, 1949. Pp. 17–25. Solution of simultaneous linear equations.

6.10. Quenouille, M. H. Associated Measurements. New York, Academic Press, Inc., 1952. Graphical analysis and short-cut approximate methods of testing.

6.11. Rider, P. R. An Introduction to Modern Statistical Methods. New York, Wiley, 1939. Pp. 27–66, 83–85, 93–98.

6.12. Scarborough, J. B. Numerical Mathematical Analysis, 2nd ed. Baltimore, The Johns Hopkins Press, 1950. Pp. 463–69, 472–74.

6.13. Snedecor, G. W. Statistical Methods, 4th ed. Ames, Iowa, Iowa State College Press, 1946. Pp. 103–67, 318–72.

6.14. Wallis, W. A. "Tolerance Intervals for Linear Regression," in Proceedings of the Second Berkeley Symposium on Mathematical Statistics and Probability, ed. by J. Neyman. Berkeley, University of California Press, 1951.

6.15. Youden, W. J. "Technique for Testing the Accuracy of Analytical Data," IND ENG CHEM, ANAL ED, Vol. 19 (1947), pp. 946–50.

Chapter 7

QUALITY-CONTROL CHARTS

7.0. Introduction

Quality-control charts are a graphical means for detecting systematic variation from the quality to be expected in a continuous production line; that is, variation greater than the random fluctuation which is inevitable and allowable. The charts may be used as a simple method of performing statistical tests (such as those in Chapters 2–5) of the hypothesis that subsequently produced articles have essentially the same quality characteristics as previously produced articles. Quality-control charts relate to a production line or process in action and indicate when a process should be examined for trouble; whereas acceptance sampling (Chapter 8) relates to entire lots of product, and enables a buyer to decide on the basis of a sample whether to accept a given lot from a supplier.

Both quality-control and acceptance inspection procedures are based on judgment of the whole by a part; that is, they are based on sampling. These modern techniques have been found preferable to screening (100% inspection) in most cases because of lower cost. The work load in screening sometimes requires each inspection to be superficial or haphazard, so that a sampling plan involving careful inspection may actually give more reliable information about a whole process. Also, if the test destroys the item inspected (a flash bulb, for instance), screening is out of the question.

Quality-control charts may be used to keep a process up to a specified standard or simply to keep a process stable. A separate control chart should be used for each part produced or each different operation intended to produce different results. For instance, separate charts would be used for the operation of cutting stock into lengths for fins and for the subsequent shaping operation. To spot flaws close to their source, a sample for each control chart should be taken whenever there has been a possibility of change in the process, as when the shift of workers has changed or the pressure has been increased. Thus, the flow of product is divided into sampling units called **rational subgroups,** chosen so that a change between two subgroups may give an indication of its cause.

In both quality-control work and acceptance sampling, a product may be judged by **attributes** or **variables**. The attributes method involves merely a decision as to whether an item is acceptable or defective; the variables method involves taking and recording measurements. These two approaches are familiar in assigning school grades, which may be recorded as "pass" or "fail" (attributes), or as percentages (variables).

Variables testing preserves more information from the sample, but is sometimes unnecessary or inapplicable. For instance, to test whether the length of a fuze falls between $1\frac{1}{4}$ and $1\frac{1}{2}$ inches, we may record the measured length, or we may simply record "within tolerances" or "not within tolerances"; but to test whether an aluminum disk has been inserted during assembly, we can only record "yes" or "no."

The technique of making quality-control charts is to set up, in graphical form, limits for a sample statistic on the basis of approximately 25 preliminary samples from production, all samples preferably having the same number of items. If one of the preliminary samples yields a statistic outside the computed control limits, the production was not initially under control and should be stabilized before the control chart is set up. The sample statistic may be the proportion defective, number of defective items, mean, range, etc. The values of the statistic should cluster about a **central line** (central value), which may be a set nominal value in the case of production to a specified standard, or the mean of previously obtained values in the case of production which is to be kept stable. The upper and lower control limits (UCL and LCL, respectively) are usually set up symmetrically at $\pm 3\sigma$ from the central line (where σ is the standard deviation of the statistic). Then so long as production remains satisfactory, the probability that a value outside these limits will occur is less than 0.3% (for a normal distribution). Therefore, if during the course of production a value from one of the rational subgroups should be recorded outside the limits on its control chart, it would be concluded that a change in the production process had occurred and the process would immediately be investigated to determine the cause of the change. In this way many poor production practices are caught and corrected before the production of large numbers of defectives. A typical control chart is shown in Fig. 7.1.

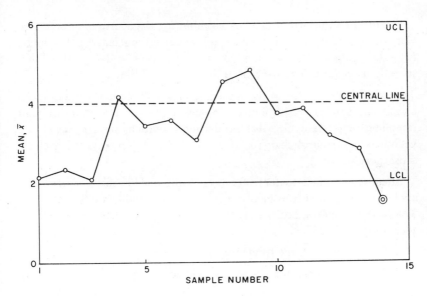

FIG. 7.1. Typical Control Chart.

A control chart showing constantly increasing (or decreasing) values may lead to an investigation before the control limits are actually passed. Values falling consistently very close to the central line should indicate excellent production, but may reflect nonrandom sampling. Values consistently below (or above) the central line may suggest trouble even when they do not pass the control limits.

The steps in controlling quality by statistical charts are the following:

1. Decide on a statistic (based on attributes or on variables) for each production operation.

2. Divide each process into rational subgroups.

3. Decide on a sample size n.

4. Construct a control chart showing the central line, UCL, and LCL for the statistic, based on at least 25 preliminary samples from satisfactory production, all samples preferably having the same number of items. The central line may have a preassigned standard value or may be determined from the preliminary samples. (Sec. 7.1 and 7.2 explain the methods of computing the limits.)

5. Plot the values from samples taken during the course of production. Consider as indications of trouble: (*a*) values falling outside the limits; (*b*) values consistently too close to the central line; and (*c*) values consistently above (or below) the central line.

6. If a situation listed in the preceding paragraph occurs: (*a*) locate and correct any flaw in production; or (*b*) investigate the possibility of poor sampling; or (*c*) recognize that production conditions have changed and institute a new control chart (if, after the change, the process still yields a satisfactory product).

Attention is called to the references at the end of this chapter. Also applicable to the problems involved in controlling quality are Sec. 4.1, Run Test for Randomness; Sec. 4.9, Tolerance Intervals; and Chapter 8, Acceptance Sampling.

In Sec. 7.1 and 7.2, the following notation is adopted:[1]

k Number of preliminary samples used to set up chart (at least 25).

n Sample size.

$\bar{x}$ Sample average, computed as $\bar{x} = \sum\limits_{i=1}^{n} x_i/n$.

S Sample sum, computed as $S = \sum\limits_{i=1}^{n} x_i$. Note that $\bar{x} = S/n$. (S can be used instead of $\bar{x}$ in a control chart. This eliminates any need for division by n.)

R Sample range, computed as largest reading minus smallest reading in a sample.

s Sample standard deviation, computed as

$$s = \sqrt{\frac{n \sum x_i^2 - (\sum x_i)^2}{n(n-1)}}$$

p Proportion defective in sample, computed as

$$p = \frac{\text{number of defective items found}}{\text{number of items inspected}}$$

[1] The formula and notation used here for the sample standard deviation s are consistent with the rest of the Manual, but inconsistent with much of the control-chart literature. In using references the reader should check to see how s or σ has been defined.

c Number of defects in sample. If several defects are found in one item, they are all counted.

$\bar{s}$ Weighted sample standard deviation obtained from k samples of slightly differing sizes, computed as

$$\bar{s} = \frac{n_1 s_1 + n_2 s_2 + \cdots + n_k s_k}{n_1 + n_2 + \cdots + n_k}$$

$\bar{s}$ reduces to $\bar{s}$ if the sample sizes n_j are equal.

NOTE: A bar over any statistic indicates an average taken over all the values of that statistic. Thus, $\bar{S} = \sum\limits_{j=1}^{k} S_j/k$, that is, the sum of the sums S_j for each of the k samples, divided by k. Similarly, $\bar{\bar{x}} = \sum\limits_{j=1}^{k} \bar{x}_j/k$, the average of the k averages $\bar{x}_j$, and $\bar{p} = \sum\limits_{j=1}^{k} p_j/k$, the average of the k sample proportions defective.

The constant factors A, A_0, A_1, A_2, B_1, B_2, B_3, B_4, c_2, D_1, D_2, D_3, D_4, and d_2 are found in Appendix Table 18.

7.1. Charts Using Attributes

This section of the Manual is presented in tabular form for compactness and ease of comparison. By using Table 7.1 the reader can readily determine the central line and the control limits for any quality-control chart of the attributes type.

Example. The average proportion defective $\bar{p}$ for 25 preliminary samples of 50 igniters each is 0.10. What are the control limits for p for samples of size 50?

The control limits are computed as

$$\bar{p} \pm 3\sqrt{\frac{\bar{p}(1-\bar{p})}{n}} = 0.10 \pm 3\sqrt{\frac{(0.10)(0.90)}{50}}$$

$$= -0.03 \text{ and } 0.23$$

Since a negative proportion defective is impossible, we replace the lower control limit, -0.03, by zero.

TABLE 7.1. GUIDE FOR CONSTRUCTING CONTROL CHARTS BASED ON ATTRIBUTES

Quantity controlled	Central line	Lower control limit (LCL)	Upper control limit (UCL)	Sample size	Remarks
Proportion defective,[a] p	$\bar{p}$	$\bar{p} - 3\sqrt{\dfrac{\bar{p}(1-\bar{p})}{n}}$ or zero	$\bar{p} + 3\sqrt{\dfrac{\bar{p}(1-\bar{p})}{n}}$	Preferably large enough to make $n\bar{p} \geq 4$	$\bar{p}$ may either be observed or specified as a standard value. If $\bar{p} < 0.05$, use a $\bar{c}$ chart with $\bar{c} = n\bar{p}$.
Number of defectives, np	$n\bar{p}$	$n\bar{p} - 3\sqrt{n\bar{p}(1-\bar{p})}$ or zero	$n\bar{p} + 3\sqrt{n\bar{p}(1-\bar{p})}$	Preferably large enough to make $n\bar{p} \geq 4$	np is the number of items found to be defective. If $\bar{p} < 0.05$, use a $\bar{c}$ chart with $\bar{c} = n\bar{p}$.
Number of defects, c	$\bar{c}$	$\bar{c} - 3\sqrt{\bar{c}}$ or zero	$\bar{c} + 3\sqrt{\bar{c}}$	Preferably large enough to make $\bar{c} \geq 4$	Expected number of defects should be small compared with maximum number of defects possible. $\bar{c}$ may either be observed or specified as a standard value.

[a] For a variables method of controlling proportion defective, see Bowker, Albert H., and Henry P. Goode, *Sampling Inspection by Variables*, New York, McGraw-Hill, 1952, pp. 109–13.

These limits, along with the central line and the first 12 subsequent sample values of p, are shown in Fig. 7.2. Samples 4 through 12 show constantly increasing p. Since such a long unbroken trend is unlikely to result from chance alone, the production process may well be checked at this point before production continues.

Example. In recent satisfactory production, an average of 4.1 flaws has been found for each sample of 20 propellant grains. For instance, the first sample of 20 grains contained one flaw in the eleventh grain ($c = 1$); the second sample of 20 grains contained two flaws in the third grain and one flaw in the seventeenth grain ($c = 3$). Construct a 3σ-limit control chart.

The control limits are $\bar{c} \pm 3\sqrt{\bar{c}} = 4.1 \pm 3\sqrt{4.1} = 4.1 \pm 6.1$. Since c is never negative, we use LCL = 0, UCL = 10.2.

From Fig. 7.3 the production process appears to be under control.

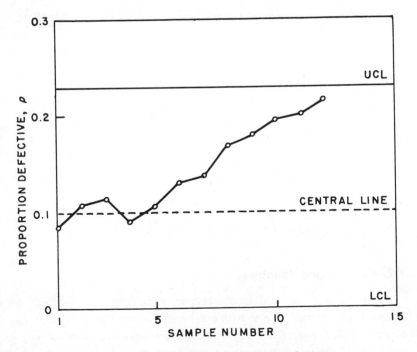

FIG. 7.2. Control Chart for Proportion Defective.

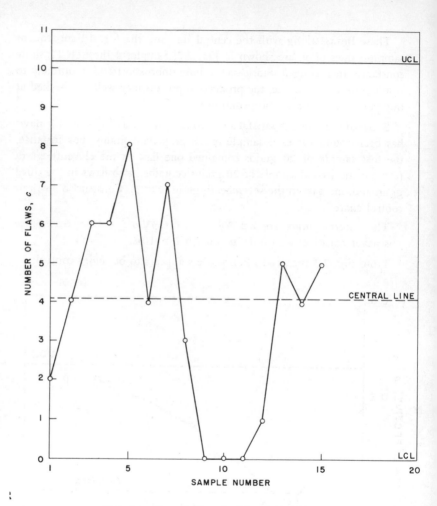

FIG. 7.3. Control Chart for Number of Flaws.

7.2. Charts Using Variables

Table 7.2 is to be used in conjunction with Appendix Table 18 for setting up the central line and control limits for any quality-control chart of the variables type. Allowance for varying sample sizes is made only in two cases. Reference 7.1 (pages 60–99) gives methods and examples for

TABLE 7.2.   GUIDE FOR CONSTRUCTING CONTROL CHARTS BASED ON VARIABLES

Quantity controlled	Central line	Lower control limit (LCL)	Upper control limit (UCL)	Sample size	Remarks
Mean $\bar{x}$	$\bar{\bar{x}}$	$\bar{\bar{x}} - A_2\bar{R}$	$\bar{\bar{x}} + A_2\bar{R}$	Small, preferably 10 or fewer	Uses range R. No standard μ and σ given.
Sum S	$\bar{S}$	$\bar{S} - A_n\bar{R}$	$\bar{S} + A_n\bar{R}$	Small, preferably 10 or fewer	No standard μ and σ given.
Mean $\bar{x}$	$\bar{\bar{x}}$	$\bar{\bar{x}} - \bar{s}A_1\sqrt{\frac{n-1}{n}}$	$\bar{\bar{x}} + \bar{s}A_1\sqrt{\frac{n-1}{n}}$	25 or fewer, constant size	No standard μ and σ given.
Mean $\bar{x}$	$\bar{\bar{x}}$	$\bar{\bar{x}} - 3\bar{s}/\sqrt{\bar{n}}$	$\bar{\bar{x}} + 3\bar{s}/\sqrt{\bar{n}}$	Sample sizes $\geqslant 25$, may vary slightly	No standard μ and σ given.
Mean $\bar{x}$	μ	$\mu - A\sigma$	$\mu + A\sigma$	n	μ and σ specified (standard values are to be maintained).
Range R	$\bar{R}$	$D_3\bar{R}$	$D_4\bar{R}$	Small, preferably 10 or fewer	No standard σ given.
Range R	$d_2\sigma$	$D_1\sigma$	$D_2\sigma$	Small, preferably 10 or fewer	σ specified (standard value to be maintained).
Sample standard deviation s	$\bar{s}$	$B_3\bar{s}$	$B_4\bar{s}$	25 or fewer, constant size	No standard σ given.
Sample standard deviation s	$\bar{s}$	$\bar{s} - 3\bar{s}/\sqrt{2\bar{n}}$	$\bar{s} + 3\bar{s}/\sqrt{2\bar{n}}$	Sample sizes $\geqslant 25$, may vary slightly	No standard σ given.
Sample standard deviation s	$\sigma c_2\sqrt{\frac{n}{n-1}}$	$\sigma B_1\sqrt{\frac{n}{n-1}}$	$\sigma B_2\sqrt{\frac{n}{n-1}}$	n	σ specified (standard value to be maintained).

TABLE 7.3. MEAN AND RANGE DATA IN SECONDS FOR
SAMPLES FROM FLARE PRODUCTION

Sample	Mean, $\bar{x}$	Range, R	Sample	Mean, $\bar{x}$	Range, R
1	64.97	9.8	14	66.60	0.6
2	64.60	9.8	15	66.12	6.3
3	64.12	8.4	16	63.22	7.5
4	68.52	3.9	17	62.85	6.7
5	68.35	7.6	18	62.37	4.9
6	67.87	8.7	19	61.97	6.7
7	64.97	0.1	20	61.60	9.9
8	64.60	9.7	21	61.12	6.9
9	64.12	7.7	22	65.72	0.1
10	63.22	7.5	23	65.35	8.3
11	62.85	1.2	24	64.87	5.2
12	62.37	9.8	25	61.97	3.2
13	66.97	6.4			

these and other cases of varying sample sizes. The average sample size may sometimes be used when variation is slight (Ref. 7.1, p. 116).

Example. Twenty-five samples of three are taken from production of flares for the purpose of establishing 3σ control limits for the time of the illumination. The times of the first sample are, for example, 60.5, 70.3, and 64.1 seconds, giving a mean of 64.97 seconds and a range of $70.3 - 60.5 = 9.8$ seconds. Sample data are given in Table 7.3.

We compute

$$\sum \bar{x} = 1{,}611.29 \qquad \sum R = 156.9$$
$$\bar{\bar{x}} = \quad 64.452 \qquad \bar{R} = \quad 6.28$$
$$n = \quad 3$$

From Appendix Table 18, $A_2 = 1.023$, $D_3 = 0$, and $D_4 = 2.575$.

Control limits for the mean $\bar{x}$ are

$$64.452 \pm (1.023)(6.28) = \begin{cases} 70.88 = \text{UCL} \\ 58.03 = \text{LCL} \end{cases}$$

Control limits for the range R are

$$\begin{cases} D_3\bar{R} = (0)\,(6.28) = 0 \\ D_4\bar{R} = (2.575)(6.28) = 16.17 \end{cases}$$

Figures 7.4 and 7.5 are control charts using these limits.

BIBLIOGRAPHY

7.1. American Society for Testing Materials, Committee E-11. ASTM Manual on Quality Control of Materials. Philadelphia, ASTM, January 1951. This is a thorough and handy manual for the common types of quality control listed in Sec. 7.1 and 7.2. Many detailed examples are given.

7.2. Bureau of Ordnance, Department of the Navy. An Introduction to Statistical Quality Control. Washington, GPO (undated). This pamphlet is on a very elementary level and is clear and well illustrated. It shows a good example of the savings made by controlling quality.

7.3. ———. The Quality Control Technique. Washington, BuOrd, 26 August 1949. (NAVORD Report 492.) This report explains how to construct $\bar{x}$ and R charts and gives sample control charts.

7.4. Grant, E. L. Statistical Quality Control, 2nd ed. New York, McGraw-Hill, 1952.

7.5. Hotelling, H. "Multivariate Quality Control, Illustrated by the Air Testing of Sample Bombsights," in Selected Techniques of Statistical Analysis, ed. by Churchill Eisenhart, M. W. Hastay, and W. A. Wallis. New York, McGraw-Hill, 1947. Inspection plans based on a generalized t statistic are developed for products assembled from several component parts.

7.6. Juran, J. M. Quality-Control Handbook. New York, McGraw-Hill, 1952.

7.7. Peach, P. Industrial Statistics and Quality Control, 2nd ed. Raleigh, N. C., Edwards & Broughton Co., 1947. The book includes special material on operations subject to trends and on quality control for short-run production.

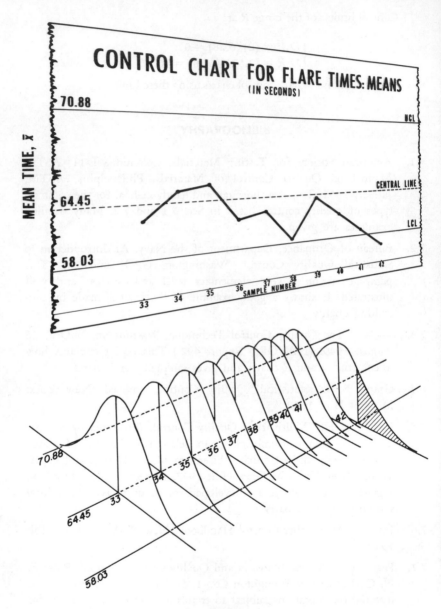

FIG. 7.4. Mean Out of Control.

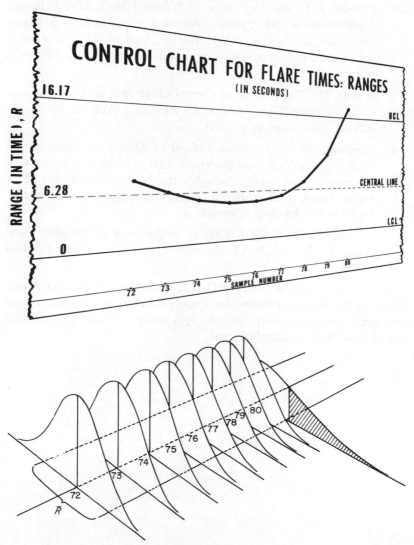

FIG. 7.5. Range Out of Control.

7.8.   Pearson, E. S. The Application of Statistical Methods to Industrial Standardization and Quality Control. London, British Standards Institution, November 1935. (British Standard 600.) The book explains the definition and theory behind quality control, presenting many examples.

7.9.   Scheffé, H. "The Relation of Control Charts to Analysis of Variance and Chi-Square Tests," AM STAT ASSOC, J, Vol. 42 (1947), pp. 425–31 (corrections on p. 634).

7.10. Shewhart, W. A. Economic Control of Quality of Manufactured Product. New York, Van Nostrand, 1931. The book includes many illustrations and treats thoroughly the presentation of data. It is the first book to be published on statistical quality control and continues to be a standard reference.

7.11. Tippett, L. H. C. Technological Applications of Statistics. New York, Wiley, 1950. Part I covers the construction and use of control charts.

The monthly journal, INDUSTRIAL QUALITY CONTROL, published by the American Society for Quality Control, contains material on the applications of quality control, with many accounts of difficulties encountered and successes achieved.

Chapter 8

ACCEPTANCE SAMPLING

8.0. Introduction

The use of statistical charts by a producer in controlling quality during production was explained in Chapter 7. This chapter will deal with **acceptance sampling**; that is, with methods by which the buyer may discriminate between acceptable and unacceptable lots of product. The methods also produce incidental information that can be useful in the control of quality. (So-called "continuous sampling" does not require division of the product into lots, but uses samples drawn directly from the production line. See Ref. 8.4.)

If a buyer finds that it is too expensive, too difficult, or impossible to screen a production lot (that is, to inspect it 100%), he may decide from a sample whether to accept or to reject the entire lot. A Navy inspector may decide on the basis of a sample from each lot of fuzes delivered by a contractor whether to accept the production lot.

Acceptance sampling may be based on either **attributes** or **variables**. An inspector checking fuzes for delay times, for instance, might judge each fuze as satisfactory or defective (attributes), or he might record the delay time for each sample fuze and compute a statistic for deciding whether to accept the whole lot (variables). (See also Sec. 7.0.)

In the attributes case, the lot is rejected if the sample contains too many defectives. In the variables case, the criterion may be one-sided or two-sided, depending upon specifications. If, for example, the specifications set an upper limit U on the acceptable delay time, the inspector computes the statistic $(U - \bar{x})/s + 10$ from the delay times of the sample and rejects the lot if this quantity is less than an acceptance constant A_U; otherwise he accepts the lot. If, instead, the specifications set a lower limit L, the inspector rejects the lot if the statistic $(L - \bar{x})/s + 10$ exceeds an acceptance constant A_L; otherwise he accepts the lot. In case the specifications set both upper and lower limits on the delay time, the inspector should use a two-sided criterion rather than two one-sided criteria. References 8.2 and 8.3 provide slightly different plans for this case, that of Ref. 8.2 being the more accurate and that of Ref. 8.3 being the more easily applied.

Variables inspection may provide for bounds on $\bar{x}$ alone or on s alone, instead of on a combination. For these techniques, see Ref. 8.1, pages 79–116.

The plans commonly used by naval ordnance inspectors for acceptance sampling by attributes are given in Ref. 8.5; those for acceptance sampling by variables, in Ref. 8.3. A collection of variables plans is given also in Ref. 8.2. This chapter will acquaint the reader with the plans given in Ref. 8.2, 8.3, and 8.5, and explain briefly the construction of such plans.

8.1. Attributes or Variables

The decision whether to use attributes or to use variables may be made on the basis of over-all cost, keeping in mind the following points:

1. A larger sample is required for attributes sampling than for variables sampling to obtain equivalent discrimination between good and bad lots. Therefore, variables inspection is preferred if sample items are costly and inspection is destructive. This consideration is important, for instance, in the case of a weapon that is expensive or in short supply.

2. The actual measurements and computations required for variables inspection may be more costly than the yes-or-no decision and tallying required for attributes testing. This must be taken into account whenever the testing itself is difficult, expensive, or time-consuming.

3. Variables methods produce as a by-product information that can be valuable in diagnosing production ills.

4. Variables plans depend for exactness on an assumption of normality in the distribution of the variable measured, though the plans may be used as approximate methods when the distribution departs from normality. Attributes plans are not subject to such a restriction. (See Sec. 4.3 for tests of normality.)

5. Attributes sampling is more widely known than variables sampling and therefore may require less training of inspectors.

8.2. Choice of Plan

There are many possible bases for choosing one sampling plan rather than another. In some cases a physical situation or the limitations of a budget may dictate the sample size or other aspects of the plan. The methods given here, which follow those of Ref. 8.2, 8.3, and 8.5, base the choice of a plan on setting the risks of rejecting a good lot or of accepting a bad one.

Having decided what is to constitute an **inspection lot** of the product (this may be a **production lot**, part of a production lot, or several production lots taken together), what is to constitute a **unit** or **item** in the lot, and whether to test by attributes or by variables, we must decide on (1) an acceptable-quality level, (2) the number of groups of items to be sampled, and (3) the inspection level. These factors will be dealt with in the sections that follow.

8.2.1. Acceptable-Quality Level (AQL)

Each sampling plan has the property that the more defectives a lot contains the more likely it is to be rejected. Since the inspection is not 100%, however, lots with a few defectives will often be accepted. The **AQL** is the percentage of defective items in a lot that we are willing to tolerate most of the time. The plans in Ref. 8.3 are arranged so that, on the average, lots with AQL% defectives will be accepted from 90% of the time for small lots to 99% for large lots; the plans in Ref. 8.5 will accept lots having AQL% defective from 80% of the time for small lots to 99.8% for large lots; and the plans in Ref. 8.2 will accept lots having AQL% defective 95% of the time.

In Ref. 8.5, a range of AQL values is approximated by one value in that range. For instance, to apply the plans for any specified AQL from 0.040 to 0.069%, we use the tabled value for AQL 0.065%. (See Table I in Ref. 8.5 for these representative AQL's.) In Ref. 8.5, AQL's over 10.0% refer only to the number of defects per hundred items (allowing several defects per item), rather than to the percentage of defective items in the lot. Reference 8.3 gives plans for six AQL's: 0.10, 0.25, 1.0, 2.5, 4.0, and 6.5%. Reference 8.2 gives plans for 15 ranges of AQL.

8.2.2. Single, Double, or Multiple Sampling

A buyer may make his decision to accept or reject a lot on the basis of one sample; that is, he may use a **single-sampling plan**. Or he may use a **double-sampling plan**, under which he draws a first sample and then makes one of the following decisions. If the first sample is sufficiently good, he will accept the lot without testing further; if it is sufficiently poor, he will reject the lot; if it is intermediate, he will inspect a second sample before deciding. A **multiple-sampling plan** extends the same idea to several samples drawn in sequence from the same lot, with the possibility of a decision on the lot (without further sampling) at each stage.

An attributes plan specifies an acceptance number Ac and a rejection number Re for each step of sampling. If the sample has Ac or fewer defective items, the buyer accepts the lot without further sampling. If the sample has Re or more defective items, he rejects the lot without further sampling. If the number of defectives is between Ac and Re, he draws another sample before deciding. After each successive sampling, he makes a decision on the basis of the total sample accumulated, using the acceptance and rejection numbers for that stage. For the last possible sample in each plan (the only sample for a single-sampling plan, the second sample for a double-sampling plan, and the final sample for a multiple-sampling plan) we have $Re = Ac + 1$, so that the lot will either be accepted (Ac or fewer defective sample items) or rejected (Re or more defectives) with no further sampling.

For variables inspection, only single- and double-sampling plans are available; see Ref. 8.2 and 8.3. To carry out, for example, a double-sampling plan, the buyer makes use of constants k_a, k_r, and k_t given in Ref. 8.2. He computes for the first sample the mean

$$\bar{x}_1 = \frac{\sum x_1}{n_1}$$

and standard deviation

$$s_1 = \sqrt{\frac{n_1 \sum x_1^2 - (\sum x_1)^2}{n_1 (n_1 - 1)}}$$

If $\bar{x}_1 + k_a s_1 \leqslant U$ (upper limit), he accepts the lot. If $\bar{x}_1 + k_r s_1 \geqslant U$, he rejects the lot. If neither decision can be made, he draws a second sample.

Computing

$$\bar{x}_t = \frac{\sum x_1 + \sum x_2}{n_1 + n_2}$$

and

$$s_t = \sqrt{\frac{(n_1 + n_2)\left(\sum x_1^2 + \sum x_2^2\right) - \left(\sum x_1 + \sum x_2\right)^2}{(n_1 + n_2)(n_1 + n_2 - 1)}}$$

for the total sample, he accepts the lot if $\bar{x}_t + k_t s_t \leqslant U$ and rejects it if $\bar{x}_t + k_t s_t > U$.

In contrast to the single-sampling plan, a plan that allows several samples does not specify in advance how many items will be required, since the number depends on whether one of the early samples yields a decision. However, the maximum number of items that could possibly be needed for the total sample is specified by the plan. The average sample number for a given lot quality can be computed (with some difficulty) by formulas given in Ref. 8.9, pp. 203–11, for double or multiple sampling by attributes. For double sampling by variables, the average sample number can be computed by formulas given in Ref. 8.2, p. 128. This average sample number is usually smaller than the fixed sample size of a single-sampling plan, because a multiple plan gives the buyer a chance to make his decision on the basis of one of the relatively small early samples if it happens to be poor enough or good enough.

8.2.3. Inspection Level

After a decision has been made to use a single-, double-, or multiple-sampling plan, an **inspection level** must be chosen. Three levels of inspection, designated I, II, and III, correspond to the different levels of the importance of detecting defective items at the expense of large sample sizes. Inspection level III is appropriate if defectives must be rejected whenever possible, regardless of the size of the sample, as might be the case when safety from blowups is a consideration. If, on the other hand, the cost of testing is unusually high and the acceptance of some defectives is not a serious matter, inspection level I is appropriate. Level II, a compromise, should be used unless there is a special need for one of the other levels.

The choice of an inspection level and of the entire sampling plan should be made only after careful consideration of the risks of error involved, the cost of inspection, and the costs associated with incorrect decisions. The risks of error of a particular plan may be obtained from an operating-characteristic (OC) curve, as described in Sec. 8.3 below. Special problems, for instance those involving safety, may require construction of special plans.

8.2.4. Sample Size and Severity of Inspection

Table III of Ref. 8.5, Table III of Ref. 8.3, or Table A of Ref. 8.2 will yield the **sample-size code letter** for a given lot size. This code letter is required when entering later tables for the characteristics of the desired sampling plan, in particular the sample size.

The **severity** of inspection influences the code letter and hence the characteristics of the plan itself. Unless previous inspection has indicated otherwise, **normal** inspection is used. References 8.2, 8.3, and 8.5 provide also for **reduced** inspection and **tightened** inspection at each level, and the criteria for their use. The standard of inspection may be reduced when the quality of production has been uniformly good over a period of time, or tightened when the quality has been poor.

Example. For an attributes plan for a lot of size 100, we find from Table III of Ref. 8.5 the code letter F for normal inspection at level II. For a variables plan, we find from Table III of Ref. 8.3 the code letter B for normal inspection at level II.

If the sample size n is a constant proportion of the lot size N, the cost of inspection per unit produced remains constant; however, the risks of wrong decisions are not constant, but increase with decreasing lot size. In the plans of Ref. 8.2, 8.3, and 8.5, therefore, the sample sizes are not decreased in proportion to the lot sizes, but are given compromise values to keep the risks of rejecting good lots or of accepting bad lots from rising inordinately as the lot sizes are decreased. In acceptance sampling demanding particular attention to the risks, as in the case of safety considerations, it may be preferable to choose the sample size n (in single sampling) according to the approximation

$$n = \frac{N n_\infty}{N + n_\infty}$$

where N is the lot size and n_∞ is the sample size that would give the specified risks for an infinitely large lot size.

8.2.5. Plans for Normal Inspection

The normal inspection plan corresponding to the specified AQL and sample-size code letter can now be found as follows.

a. Attributes. For the acceptance and rejection numbers of an attributes-sampling plan, turn to Ref. 8.5, and enter Table IV-A for a single-sampling plan; Table IV-B for a double-sampling plan; and Table IV-C for a multiple-sampling plan.

Example. A buyer needs an attributes double-sampling plan to test lots of 7,000 items each. He is willing to accept, most of the time, lots that contain 0.30% defective items. What is the plan recommended by Ref. 8.5 for such a case?

From Table I of Ref. 8.5, for a specified AQL falling between 0.280 and 0.439% the representative AQL value is 0.40%. From Table III of Ref. 8.5, for a lot size of 3,201 to 8,000 the sample-size code letter is M for inspection level II. For convenience, part of Table IV-B (double-sampling plan) from Ref. 8.5 is reproduced as Table 8.1, and the appropriate figures have been italicized.

TABLE 8.1. EXCERPT FROM TABLE IV-B OF REF. 8.5

Sample-size code letter	Sample	Sample size	Cumulative sample size	AQL for normal inspection					
				0.25%		0.40%		0.65%	
				Ac	Re	Ac	Re	Ac	Re
M	First	150	150	1	3	2	5	2	7
	Second	300	450	2	3	4	5	6	7

The buyer is instructed to draw a random sample of 150 items from the lot to be tested. If among the 150 he finds 0, 1, or 2 defective items, he should accept the lot. If he finds 5, 6, 7, or more defective items, he should reject the lot. If he finds 3 or 4 defective items in the first sample, he must draw a second sample in order to reach a decision. The second sample should contain 300 items, making a total of 450. If the number of defectives from the first sample plus the number from the second is

4 or fewer, the buyer should accept the lot. If the total number of defectives is 5 or more, he should reject the lot. Notice that he will either accept or reject at this stage, if he has not done so before, so that no further sampling is required.

b. Variables. To get the acceptance constant A_U or A_L of a variables-sampling plan, enter Table IV in Ref. 8.3 (which deals with single-sampling plans only) with the desired AQL value and the sample-size code letter.

In operation, the single-sampling plan calls for calculating from the sample

$$\bar{x} = \frac{\sum_{i=1}^{n} x_i}{n}$$

and

$$s = \sqrt{\frac{n\sum x_i^2 - (\sum x_i)^2}{n(n-1)}}$$

If U is the specified upper limit of the measurement x for a nondefective item, compute the lot-quality index

$$C_U = \frac{U - \bar{x}}{s} + 10$$

and accept the lot if $C_U \geqslant A_U$ or reject the lot if $C_U < A_U$. For a lower specification limit L, compute the lot-quality index

$$C_L = \frac{L - \bar{x}}{s} + 10$$

and accept the lot if $C_L \leqslant A_L$ or reject the lot if $C_L > A_L$.

Example. Suppose that the Navy buys explosive charges from a contractor. Since a charge having a heat of explosion less than 400 calories per gram is considered acceptable, the buyer is willing to accept, most of the time, lots in which not more than 0.25% of the charges have heats of explosion greater than 400 calories per gram. The charges are easy to test, and it is imperative that faulty charges be discovered. There are 2,000 charges in each lot. What is the single-sampling plan recommended by Ref. 8.3?

From the statement of the problem, the upper specification limit U is 400 calories per gram, the AQL is 0.25%, and the inspection level is III.

From Table III of Ref. 8.3, the sample-size code letter is F for normal inspection at level III for a lot of size 1,301 to 8,000. Part of Table IV of Ref. 8.3 is reproduced as Table 8.2. From this table we find that for AQL 0.25% the acceptance constant A_U is 12.45.

TABLE 8.2. EXCERPT FROM TABLE IV OF REF. 8.3

Sample-size code letter	Sample size	AQL for normal or reduced inspection		
		0.10%	0.25%	1.0%
		A_U	A_U	A_U
F	130	12.70	12.45	12.01

To dispose of each lot of 2,000 charges, the buyer draws a random sample of 130. He measures the heat of explosion x_i for each one, and calculates

$$\bar{x} = \frac{1}{130} \sum_{i=1}^{130} x_i$$

and

$$s = \sqrt{\frac{130\sum x_i^2 - (\sum x_i)^2}{(130)(129)}}$$

He computes

$$C_U = \frac{400 - \bar{x}}{s} + 10$$

and compares it with the acceptance constant $A_U = 12.45$, accepting the lot if $C_U \geqslant 12.45$, and rejecting the lot if $C_U < 12.45$.

8.3. Operating-Characteristic (OC) Curve and Its Use in Designing Plans

Since the plans described in this chapter involve sampling, there is a risk of error. References 8.2, 8.3, and 8.5 give operating-characteristic (OC) curves for the plans they include. These OC curves should be considered carefully, because the method for determining the plan to use does not fully describe the plan itself, especially with respect to the buyer's risk of accepting lots he considers definitely unsatisfactory. The OC curve shows for each possible percentage of defective items in a lot the probability of accepting a lot of that quality. The ideal plan would have a steep curve, which would discriminate sharply between good lots and bad lots. (This is a special use for the OC curve described in Sec. 1.1.9. Examples of such curves appear in Appendix Charts V, VI, VII, and VIII.)

One important quantity available from the OC curve is the **producer's risk** α. For the lot proportion defective p_1 specified by the AQL, the curve gives the probability of accepting a lot of that quality. This probability is $1 - \alpha$. (See Fig. 8.1.) For the plans of Ref. 8.2 the producer's risk is 5%, so that each curve passes through the point (AQL, 0.95). As mentioned in Sec. 8.2.1, however, this is not true for the plans of Ref. 8.3 and 8.5.

It is also important to check the OC curve for high proportions of defective items; for instance, the point (p_2, β), where p_2 is a definitely unsatisfactory proportion defective and β is the **consumer's risk** of accepting a lot of that quality. A commonly used pair of coordinates is (LTPD, 0.10), where LTPD stands for **lot tolerance percent defective** and 10% is a conventional choice for the consumer's risk.

Peach (Ref. 8.8) gives a method for designing a single- or a double-sampling plan by attributes with an OC curve that passes near the two given points $(p_1, 0.95)$ and $(p_2, 0.05)$. Following this method, with $p_1 = \text{AQL}$, form the operating ratio

$$R_0 = \frac{p_2}{p_1}$$

For a *single-sampling plan*, enter Appendix Table 19(*a*), using R_0 or the next larger tabled value to obtain the acceptance number Ac and np_1. Divide np_1 by p_1 to get the sample size n. ($Re = Ac + 1$.)

To determine the actual abscissas p_1^* and p_2^* for the ordinates 0.95 and 0.05 of the OC curve of this plan, let $c = Ac$, and compute

$$p_1^* = \frac{c + 1}{c + 1 + (n - c) F_1}$$

$$p_2^* = \frac{(c + 1) F_2}{n - c + (c + 1) F_2}$$

where F_1 is the 5% point of the F distribution with $f_1 = 2 (n - c)$ and $f_2 = 2 (c + 1)$ degrees of freedom, and F_2 is the 5% point of the F dis-

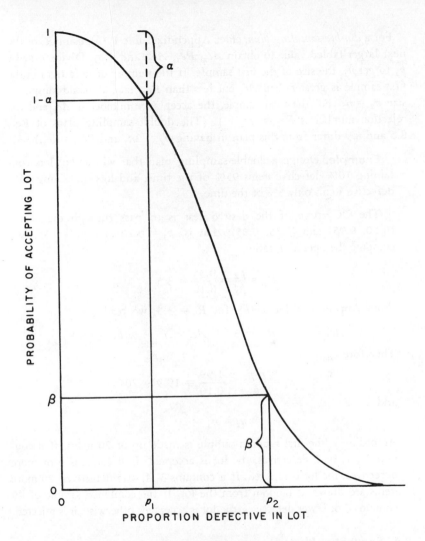

FIG. 8.1. Typical Operating-Characteristic Curve.

tribution with $f_1 = 2(c + 1)$ and $f_2 = 2(n - c)$ degrees of freedom. (See Appendix Table 5.) If these actual values do not come close enough to the desired p_1 and p_2, they can be adjusted by changes in Ac and n. Since np_1 remains nearly constant, an increase in n produces a decrease in p_1^*.

For a *double-sampling plan*, enter Appendix Table 19(b), using R_0 or the next larger tabled value to obtain Ac_1, Re_1, Ac_2, and n_1p_1. Divide n_1p_1 by p_1 to get n_1, the size of the first sample. If the number of defectives in the first sample is greater than Ac_1 but less than Re_1, take a second sample of size $n_2 = n_1$. For the total sample, the acceptance number is Ac_2 and the rejection number is $Re_2 = Ac_2 + 1$. (The double-sampling plans of Ref. 8.5 and 8.9 differ from this plan in having $n_2 = 2n_1$ and $Re_2 = Re_1$.)

Example. Design a double-sampling plan that will accept lots containing 10% defective items 95% of the time, and lots containing 35% defective items only 5% of the time.

The OC curve of the desired test is to pass through the points (0.10, 0.95) and (0.35, 0.05); that is, $p_1 = 0.10$ and $p_2 = 0.35$. We compute the operating ratio

$$R_0 = \frac{p_2}{p_1} = \frac{0.35}{0.10} = 3.5$$

From Appendix Table 19(b), for $R_0 = 3.5$, we read

$$Ac_1 = 2 \qquad Re_1 = 6 \qquad Ac_2 = 7 \qquad n_1p_1 = 1.99$$

Therefore

$$n_1 = \frac{n_1p_1}{p_1} = \frac{1.99}{0.10} = 19.9 \sim 20$$

and

$$n_2 = n_1 = 20$$

Accordingly, the first random sample is made up of 20 items. If it contains 0, 1, or 2 defectives, the lot is accepted. If it contains 6 or more defectives, the lot is rejected. If it contains 3, 4, or 5 defectives, 20 more items are drawn at random from the lot. If the combined sample of 40 contains 7 or fewer defectives, the lot is accepted; otherwise, it is rejected.

8.4. Sequential Plans

In double- and multiple-sampling plans, we take advantage of the fact that a very poor lot or a very good lot can be expected to reveal its character in a small first sample; usually, more extensive sampling is necessary only for lots of medium quality. This idea has been exploited still further in **sequential analysis**, which calls for drawing one sample item at a time with the possibility of a decision on the lot after each drawing. (Actually, the method can also be used if items are in groups, as in multiple sampling. In this case, the average increase in sample size over pure sequential sampling is no greater than the group size.)

To perform a sequential test for, say, acceptance sampling by attributes, the inspector needs a chart similar to Fig. 8.2, on which there are two slanting parallel lines. The abscissa represents the number of items drawn and the ordinate represents the number of defective items drawn. As the inspector draws and tests each item, he plots a point on the chart one unit to the right if the item is not defective; one unit to the right and one unit up if the item is defective. If a continuing path connecting all the points crosses the upper parallel line, the inspector will reject the lot. If it crosses the lower parallel line, he will accept the lot. If it remains between the two lines, he will draw another sample item.

This technique is economical in the use of sample items, is easy to apply, and is adaptable to the following:

1. Acceptance sampling using attributes (Sec. 8.4.1)
2. Making a decision, based on attributes, between *two* methods or products
3. Making one-sided and two-sided significance tests for the mean of a normal distribution
4. Making a one-sided significance test for the standard deviation of a normal distribution (Sec. 8.4.2)

Sequential analysis is *not* appropriate when it is inconvenient to wait for test results on one item at a time, or when testing is inexpensive compared with producing items for test, so that it is preferable to test all provided items in a sample of fixed size. References 8.7, 8.10, and 8.12 give clear expositions of the situations listed above, in addition to the following treatments of situations 1 and 4.

8.4.1. Sequential Plan for Proportion Defective p

To construct a chart for use in a sequential-sampling plan for proportion defective p with OC curve through the points $(p_1, 1 - \alpha)$ and (p_2, β), we draw on graph paper a horizontal axis n to represent the number of items drawn, and a vertical axis d to represent the number of defective items drawn. On these axes we draw the parallel lines

$$d_1 = -h_1 + sn \qquad \text{(lower)}$$

$$d_2 = h_2 + sn \qquad \text{(upper)}$$

To find the intercepts $-h_1$ and h_2 and the common slope s, we first compute the auxiliary quantities

$$g_1 = \log \frac{p_2}{p_1} \qquad\qquad g_2 = \log \frac{1 - p_1}{1 - p_2}$$

$$a = \log \frac{1 - \beta}{\alpha} \qquad\qquad b = \log \frac{1 - \alpha}{\beta}$$

(Taking all logarithms to the base 10, we find the values of a and b in Appendix Table 20 for common values of α and β.) Then we have

$$h_1 = \frac{b}{g_1 + g_2} \qquad\qquad h_2 = \frac{a}{g_1 + g_2}$$

$$s = \frac{g_2}{g_1 + g_2}$$

Once the chart has been drawn, its use is simple. For each item inspected, the point (n, d) is plotted. This point shows how many items have been inspected and how many of them were found defective. Notice that n will increase by one each time, moving the point one unit to the right. If the item is defective, the point will also move up one unit. As the sampling continues, the series of points plotted may be connected by a continuing path. If at any point this path crosses or meets the lower of the parallel lines, we accept the lot. If it crosses or meets the upper line, we reject the lot. As long as it remains between the lines, we continue sampling.

A **truncated sequential-sampling plan** may be used to prevent the possibility of requiring a very large sample—a possibility that is present in the ordinary sequential plan, especially for borderline lots. If we agree to stop sampling when

$$n = \frac{3ab}{g_1 g_2}$$

the change in the risks α and β of the ordinary plan will be negligible, because it is rarely necessary to test this number of samples before a decision is reached. If n gets this large with no decision, we accept the lot if the vertical distance from the last point to the lower line is less than the vertical distance from the point to the upper line. Otherwise we reject the lot.

Example. Design a truncated sequential-sampling plan that will reject only 5% of the time lots containing 0.1% defective items, but will reject 90% of the time lots containing 5% defective items.

We compute $g_1 = 1.69897$ and $g_2 = 0.02185$. From Appendix Table 20 we have $a = 1.255$ and $b = 0.978$. Then

$$h_1 = \frac{0.978}{1.69897 + 0.02185} = \frac{0.978}{1.721} = 0.568$$

$$h_2 = \frac{1.255}{1.721} = 0.729$$

$$s = \frac{0.02185}{1.721} = 0.01270$$

To truncate the plan, we agree to stop sampling after

$$n = \frac{3ab}{g_1 g_2} = \frac{3(1.255)(0.978)}{(1.69897)(0.02185)} = 99.2 \sim 99$$

items have been inspected, accepting if the continuing path is closer to the acceptance (lower) line and rejecting if the path is closer to the rejection (upper) line.

Figure 8.2 shows the sampling chart used for a hypothetical sample of 99. The sampling technique led to acceptance of the lot.

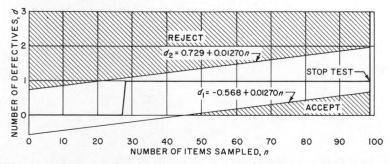

FIG. 8.2. Truncated Sequential-Sampling Chart for an Attributes Inspection Plan. The operating-characteristic curve passes through (0.001, 0.95) and (0.05, 0.10).

8.4.2. Sequential Plan for Standard Deviation σ

If the measurements of a lot or process follow a normal distribution with unknown mean μ and standard deviation σ, we can construct a sequential method for testing the null hypothesis that σ has a hypothetical value σ_1 against the alternative $\sigma = \sigma_2 > \sigma_1$. (See Sec. 3.2.2 for the non-sequential one-sided test.) Suppose there is an assigned risk α of rejecting the hypothesis when σ *does* equal σ_1, and an assigned risk β of accepting the hypothesis when actually σ has the higher value σ_2.

Let n stand for the number of sample items inspected, and let Z stand for the sum of squared deviations from the sample mean. Then

$$Z = \sum_{i=1}^{n} (x_i - \bar{x})^2 = \frac{n\sum x_i^2 - (\sum x_i)^2}{n}$$

The latter form is to be used for computing. Draw n and Z axes on ordinary rectangular graph paper. The acceptance and rejection lines are then

$$Z_1 = -h_1 + s(n-1) \qquad \text{(lower)}$$
$$Z_2 = h_2 + s(n-1) \qquad \text{(upper)}$$

respectively. To find the intercepts $-h_1$ and h_2 and the common slope s, we first compute the auxiliary quantities

$$g = 0.43429 \left(\frac{1}{\sigma_1^2} - \frac{1}{\sigma_2^2} \right)$$
$$a = \log_{10} \frac{1-\beta}{\alpha}$$
$$b = \log_{10} \frac{1-\alpha}{\beta}$$

Then we have

$$h_1 = \frac{2b}{g} \qquad h_2 = \frac{2a}{g}$$
$$s = \frac{\log_{10}(\sigma_2^2/\sigma_1^2)}{g}$$

Appendix Table 20 may be used to get a and b for common values of α and β.

The chart for this technique, shown as Fig. 8.3, should be used in the same manner as the chart for the proportion-defective test given in Sec. 8.4.1.

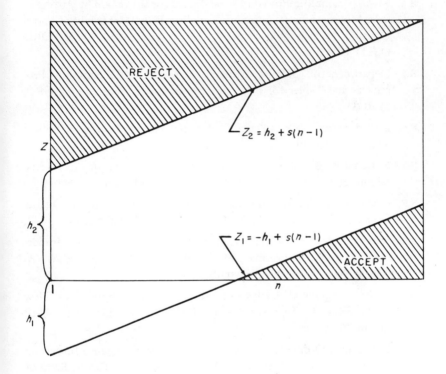

FIG. 8.3. Typical Sequential-Sampling Chart for Testing the Hypothesis That $\sigma = \sigma_1$ Against the Alternative, $\sigma = \sigma_2 > \sigma_1$.

BIBLIOGRAPHY

8.1. American Statistical Association. Acceptance Sampling. Washington, ASA, 1950.

8.2. Bowker, Albert H., and Henry P. Goode. Sampling Inspection by Variables. New York, McGraw-Hill, 1952.

8.3. Bureau of Ordnance. Naval Ordnance Standard Sampling Procedures and Tables for Inspection by Variables. Washington, BuOrd, 8 May

1952. (NAVORD OSTD 80; NAVORD OSTD 80, Change 1, 8 September 1952, 1 p.)

8.4. ————. Sampling Procedures and Tables for Inspection by Attributes on a Moving Line (Continuous Sampling Plans), by Quality Control Division. Washington, BuOrd, 15 August 1952. (NAVORD OSTD 81.)

8.5. Department of Defense. Military Standard 105A: Sampling Procedures and Tables for Inspection by Attributes. Washington, GPO, 1950.

8.6. Dodge, Harold F., and Harry G. Romig. Sampling Inspection Tables: Single and Double Sampling. New York, Wiley, 1944.

8.7. National Bureau of Standards. Tables To Facilitate Sequential t-Tests. Washington, GPO, 16 May 1951. (Applied Mathematics Series 7.)

8.8. Peach, Paul. An Introduction to Industrial Statistics and Quality Control, 2nd ed. Raleigh, N. C., Edwards & Broughton Co., 1947.

8.9. Statistical Research Group, Columbia University. Sampling Inspection, ed. by H. A. Freeman, M. Friedman, F. Mosteller, and W. A. Wallis. New York, McGraw-Hill, 1948.

8.10. ————. Sequential Analysis of Statistical Data: Applications. New York, SRG, 15 September 1945. (SRG Report 255, rev., AMP Report 30.2R, rev.)

8.11. U. S. Naval Ordnance Test Station, Inyokern. Acceptance Inspection by Variables in the Presence of Measurement Error, Part 1, Ratio of Measurement Error to Product Variability Assumed Constant, by Edwin L. Crow and Edward A. Fay. China Lake, Calif., NOTS, 27 December 1951. (NAVORD Report 1940, Part 1, NOTS 483.)

8.12. Wald, Abraham. Sequential Analysis. New York, Wiley, 1947.

APPENDIX TABLES

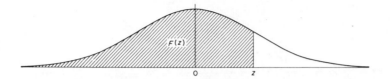

TABLE 1. Cumulative Normal Distribution *

$$F(z) = \int_{-\infty}^{z} \frac{1}{\sqrt{2\pi}} \, e^{-z^2/2} \, dz$$

z	.00	.01	.02	.03	.04	.05	.06	.07	.08	.09
.0	.5000	.5040	.5080	.5120	.5160	.5199	.5239	.5279	.5319	.5359
.1	.5398	.5438	.5478	.5517	.5557	.5596	.5636	.5675	.5714	.5753
.2	.5793	.5832	.5871	.5910	.5948	.5987	.6026	.6064	.6103	.6141
.3	.6179	.6217	.6255	.6293	.6331	.6368	.6406	.6443	.6480	.6517
.4	.6554	.6591	.6628	.6664	.6700	.6736	.6772	.6808	.6844	.6879
.5	.6915	.6950	.6985	.7019	.7054	.7088	.7123	.7157	.7190	.7224
.6	.7257	.7291	.7324	.7357	.7389	.7422	.7454	.7486	.7517	.7549
.7	.7580	.7611	.7642	.7673	.7704	.7734	.7764	.7794	.7823	.7852
.8	.7881	.7910	.7939	.7967	.7995	.8023	.8051	.8078	.8106	.8133
.9	.8159	.8186	.8212	.8238	.8264	.8289	.8315	.8340	.8365	.8389
1.0	.8413	.8438	.8461	.8485	.8508	.8531	.8554	.8577	.8599	.8621
1.1	.8643	.8665	.8686	.8708	.8729	.8749	.8770	.8790	.8810	.8830
1.2	.8849	.8869	.8888	.8907	.8925	.8944	.8962	.8980	.8997	.9015
1.3	.9032	.9049	.9066	.9082	.9099	.9115	.9131	.9147	.9162	.9177
1.4	.9192	.9207	.9222	.9236	.9251	.9265	.9279	.9292	.9306	.9319
1.5	.9332	.9345	.9357	.9370	.9382	.9394	.9406	.9418	.9429	.9441
1.6	.9452	.9463	.9474	.9484	.9495	.9505	.9515	.9525	.9535	.9545
1.7	.9554	.9564	.9573	.9582	.9591	.9599	.9608	.9616	.9625	.9633
1.8	.9641	.9649	.9656	.9664	.9671	.9678	.9686	.9693	.9699	.9706
1.9	.9713	.9719	.9726	.9732	.9738	.9744	.9750	.9756	.9761	.9767
2.0	.9772	.9778	.9783	.9788	.9793	.9798	.9803	.9808	.9812	.9817
2.1	.9821	.9826	.9830	.9834	.9838	.9842	.9846	.9850	.9854	.9857
2.2	.9861	.9864	.9868	.9871	.9875	.9878	.9881	.9884	.9887	.9890
2.3	.9893	.9896	.9898	.9901	.9904	.9906	.9909	.9911	.9913	.9916
2.4	.9918	.9920	.9922	.9925	.9927	.9929	.9931	.9932	.9934	.9936
2.5	.9938	.9940	.9941	.9943	.9945	.9946	.9948	.9949	.9951	.9952
2.6	.9953	.9955	.9956	.9957	.9959	.9960	.9961	.9962	.9963	.9964
2.7	.9965	.9966	.9967	.9968	.9969	.9970	.9971	.9972	.9973	.9974
2.8	.9974	.9975	.9976	.9977	.9977	.9978	.9979	.9979	.9980	.9981
2.9	.9981	.9982	.9982	.9983	.9984	.9984	.9985	.9985	.9986	.9986
3.0	.9987	.9987	.9987	.9988	.9988	.9989	.9989	.9989	.9990	.9990
3.1	.9990	.9991	.9991	.9991	.9992	.9992	.9992	.9992	.9993	.9993
3.2	.9993	.9993	.9994	.9994	.9994	.9994	.9994	.9995	.9995	.9995
3.3	.9995	.9995	.9995	.9996	.9996	.9996	.9996	.9996	.9996	.9997
3.4	.9997	.9997	.9997	.9997	.9997	.9997	.9997	.9997	.9997	.9998

° Use explained in Sec. 1.2.1. For more extensive tables, see National Bureau of Standards, *Tables of Normal Probability Functions*, Washington, U. S. Government Printing Office, 1953 (Applied Mathematics Series 23). Note that they show

$$\int_{-z}^{z} f(z)\, dz, \quad \text{not} \quad \int_{-\infty}^{z} f(z)\, dz$$

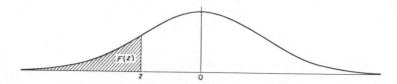

TABLE 2. PERCENTILES OF THE NORMAL DISTRIBUTION*

$$F(z) = \int_{-\infty}^{z} \frac{1}{\sqrt{2\pi}}\, e^{-z^2/2}\, dz$$

$F(z)$	z	$F(z)$	z
.0001	−3.719	.500	.000
.0005	−3.291	.550	.126
.001	−3.090	.600	.253
.005	−2.576	.650	.385
.010	−2.326		
.025	−1.960	.700	.524
.050	−1.645	.750	.674
		.800	.842
.100	−1.282	.850	1.036
.150	−1.036	.900	1.282
.200	−.842		
.250	−.674	.950	1.645
.300	−.524	.975	1.960
		.990	2.326
.350	−.385	.995	2.576
.400	−.253	.999	3.090
.450	−.126	.9995	3.291
.500	.000	.9999	3.719

* Use explained in Sec. 1.2.1. For a normally distributed variable x, we have $x = \mu + z\sigma$, where μ = mean of x and σ = standard deviation of x. For more extensive tables, see R. A. Fisher and F. Yates, *Statistical Tables*, 4th rev. ed., Edinburgh, Oliver & Boyd, Ltd., 1953, pp. 39, 60–62.

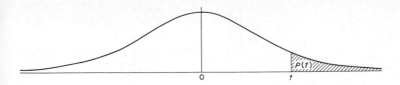

TABLE 3. Upper Percentage Points of the t Distribution *

$$P(t) = \int_t^\infty \frac{\left(\frac{f-1}{2}\right)!}{\left(\frac{f-2}{2}\right)! \sqrt{\pi f}} (1 + t^2/f)^{-(f+1)/2} dt$$

f \ $P(t)$	.40	.30	.25	.20	.15	.10	.05	.025	.01	.005	.0005
1	.325	.727	1.000	1.376	1.963	3.078	6.314	12.706	31.821	63.657	636.619
2	.289	.617	.816	1.061	1.386	1.886	2.920	4.303	6.965	9.925	31.598
3	.277	.584	.765	.978	1.250	1.638	2.353	3.182	4.541	5.841	12.924
4	.271	.569	.741	.941	1.190	1.533	2.132	2.776	3.747	4.604	8.610
5	.267	.559	.727	.920	1.156	1.476	2.015	2.571	3.365	4.032	6.869
6	.265	.553	.718	.906	1.134	1.440	1.943	2.447	3.143	3.707	5.959
7	.263	.549	.711	.896	1.119	1.415	1.895	2.365	2.998	3.499	5.408
8	.262	.546	.706	.889	1.108	1.397	1.860	2.306	2.896	3.355	5.041
9	.261	.543	.703	.883	1.100	1.383	1.833	2.262	2.821	3.250	4.781
10	.260	.542	.700	.879	1.093	1.372	1.812	2.228	2.764	3.169	4.587
11	.260	.540	.697	.876	1.088	1.363	1.796	2.201	2.718	3.106	4.437
12	.259	.539	.695	.873	1.083	1.356	1.782	2.179	2.681	3.055	4.318
13	.259	.538	.694	.870	1.079	1.350	1.771	2.160	2.650	3.012	4.221
14	.258	.537	.692	.868	1.076	1.345	1.761	2.145	2.624	2.977	4.140
15	.258	.536	.691	.866	1.074	1.341	1.753	2.131	2.602	2.947	4.073
16	.258	.535	.690	.865	1.071	1.337	1.746	2.120	2.583	2.921	4.015
17	.257	.534	.689	.863	1.069	1.333	1.740	2.110	2.567	2.898	3.965
18	.257	.534	.688	.862	1.067	1.330	1.734	2.101	2.552	2.878	3.922
19	.257	.533	.688	.861	1.066	1.328	1.729	2.093	2.539	2.861	3.883
20	.257	.533	.687	.860	1.064	1.325	1.725	2.086	2.528	2.845	3.850
21	.257	.532	.686	.859	1.063	1.323	1.721	2.080	2.518	2.831	3.819
22	.256	.532	.686	.858	1.061	1.321	1.717	2.074	2.508	2.819	3.792
23	.256	.532	.685	.858	1.060	1.319	1.714	2.069	2.500	2.807	3.767
24	.256	.531	.685	.857	1.059	1.318	1.711	2.064	2.492	2.797	3.745
25	.256	.531	.684	.856	1.058	1.316	1.708	2.060	2.485	2.787	3.725
26	.256	.531	.684	.856	1.058	1.315	1.706	2.056	2.479	2.779	3.707
27	.256	.531	.684	.855	1.057	1.314	1.703	2.052	2.473	2.771	3.690
28	.256	.530	.683	.855	1.056	1.313	1.701	2.048	2.467	2.763	3.674
29	.256	.530	.683	.854	1.055	1.311	1.699	2.045	2.462	2.756	3.659
30	.256	.530	.683	.854	1.055	1.310	1.697	2.042	2.457	2.750	3.646
40	.255	.529	.681	.851	1.050	1.303	1.684	2.021	2.423	2.704	3.551
60	.254	.527	.679	.848	1.046	1.296	1.671	2.000	2.390	2.660	3.460
120	.254	.526	.677	.845	1.041	1.289	1.658	1.980	2.358	2.617	3.373
∞	.253	.524	.674	.842	1.036	1.282	1.645	1.960	2.326	2.576	3.291

* Use explained in Sec. 2.2.2. For a two-sided (equal-tails) test the significance levels are twice the above column headings. The reciprocals of the numbers of degrees of freedom, rather than the numbers themselves, should be used for linear interpolation. To calculate the upper 0.5% point for $f = 34$, use

$$2.704 + \frac{(1/34) - (1/40)}{(1/30) - (1/40)} \ (0.046) = 2.728$$

which is correct to 3 decimal places, whereas ordinary linear interpolation would give 2.732. This table was adapted, with the permission of the authors and the publishers, from R. A. Fisher and F. Yates, *Statistical Tables*, 4th rev. ed., Edinburgh, Oliver & Boyd, Ltd., 1953, p. 40. (The original table is given in terms of the two-sided test probabilities.)

TABLE 4. UPPER PERCENTAGE POINTS OF THE χ^2 DISTRIBUTION *

$$P(\chi^2) = \int_{\chi^2}^{\infty} \frac{1}{\left(\frac{f-2}{2}\right)!} \frac{(\chi^2)^{(f-2)/2} e^{-\chi^2/2} \, d(\chi^2)}{2^{f/2}}.$$

f \ $P(\chi^2)$	.995	.990	.975	.950	.900	.750	.500	.250	.100	.050	.025	.010	.005
1	3927×10^{-8}	1571×10^{-7}	9821×10^{-7}	3932×10^{-6}	0.01579	0.1015	0.4549	1.323	2.706	3.841	5.024	6.635	7.879
2	0.01003	0.02010	0.05064	0.1026	.2107	.5754	1.386	2.773	4.605	5.991	7.378	9.210	10.60
3	.07172	.1148	.2158	.3518	.5844	1.213	2.366	4.108	6.251	7.815	9.348	11.34	12.84
4	.2070	.2971	.4844	.7107	1.064	1.923	3.357	5.385	7.779	9.488	11.14	13.28	14.86
5	.4117	.5543	.8312	1.145	1.610	2.675	4.351	6.626	9.236	11.07	12.83	15.09	16.75
6	.6757	.8721	1.237	1.635	2.204	3.455	5.348	7.841	10.64	12.59	14.45	16.81	18.55
7	.9893	1.239	1.690	2.167	2.833	4.255	6.346	9.037	12.02	14.07	16.01	18.48	20.28
8	1.344	1.646	2.180	2.733	3.490	5.071	7.344	10.22	13.36	15.51	17.53	20.09	21.96
9	1.735	2.088	2.700	3.325	4.168	5.899	8.343	11.39	14.68	16.92	19.02	21.67	23.59
10	2.156	2.558	3.247	3.940	4.865	6.737	9.342	12.55	15.99	18.31	20.48	23.21	25.19
11	2.603	3.053	3.816	4.575	5.578	7.584	10.34	13.70	17.28	19.68	21.92	24.72	26.76
12	3.074	3.571	4.404	5.226	6.304	8.438	11.34	14.85	18.55	21.03	23.34	26.22	28.30
13	3.565	4.107	5.009	5.892	7.042	9.299	12.34	15.98	19.81	22.36	24.74	27.69	29.82
14	4.075	4.660	5.629	6.571	7.790	10.17	13.34	17.12	21.06	23.68	26.12	29.14	31.32
15	4.601	5.229	6.262	7.261	8.547	11.04	14.34	18.25	22.31	25.00	27.49	30.58	32.80
16	5.142	5.812	6.908	7.962	9.312	11.91	15.34	19.37	23.54	26.30	28.85	32.00	34.27
17	5.697	6.408	7.564	8.672	10.09	12.79	16.34	20.49	24.77	27.59	30.19	33.41	35.72
18	6.265	7.015	8.231	9.390	10.86	13.68	17.34	21.60	25.99	28.87	31.53	34.81	37.16
19	6.844	7.633	8.907	10.12	11.65	14.56	18.34	22.72	27.20	30.14	32.85	36.19	38.58
20	7.434	8.260	9.591	10.85	12.44	15.45	19.34	23.83	28.41	31.41	34.17	37.57	40.00

$P(\chi^2)$ / f	.995	.990	.975	.950	.900	.750	.500	.250	.100	.050	.025	.010	.005
21	8.034	8.897	10.28	11.59	13.24	16.34	20.34	24.93	29.62	32.67	35.48	38.93	41.40
22	8.643	9.542	10.98	12.34	14.04	17.24	21.34	26.04	30.81	33.92	36.78	40.29	42.80
23	9.260	10.20	11.69	13.09	14.85	18.14	22.34	27.14	32.01	35.17	38.08	41.64	44.18
24	9.886	10.86	12.40	13.85	15.66	19.04	23.34	28.24	33.20	36.42	39.36	42.98	45.56
25	10.52	11.52	13.12	14.61	16.47	19.94	24.34	29.34	34.38	37.65	40.65	44.31	46.93
26	11.16	12.20	13.84	15.38	17.29	20.84	25.34	30.43	35.56	38.89	41.92	45.64	48.29
27	11.81	12.88	14.57	16.15	18.11	21.75	26.34	31.53	36.74	40.11	43.19	46.96	49.64
28	12.46	13.56	15.31	16.93	18.94	22.66	27.34	32.62	37.92	41.34	44.46	48.28	50.99
29	13.12	14.26	16.05	17.71	19.77	23.57	28.34	33.71	39.09	42.56	45.72	49.59	52.34
30	13.79	14.95	16.79	18.49	20.60	24.48	29.34	34.80	40.26	43.77	46.98	50.89	53.67
40	20.71	22.16	24.43	26.51	29.05	33.66	39.34	45.62	51.80	55.76	59.34	63.69	66.77
50	27.99	29.71	32.36	34.76	37.69	42.94	49.33	56.33	63.17	67.50	71.42	76.15	79.49
60	35.53	37.48	40.48	43.19	46.46	52.29	59.33	66.98	74.40	79.08	83.30	88.38	91.95
70	43.28	45.44	48.76	51.74	55.33	61.70	69.33	77.58	85.53	90.53	95.02	100.42	104.22
80	51.17	53.54	57.15	60.39	64.28	71.14	79.33	88.13	96.58	101.88	106.63	112.33	116.32
90	59.20	61.75	65.65	69.13	73.29	80.62	89.33	98.65	107.56	113.14	118.14	124.12	128.30
100	67.33	70.06	74.22	77.93	82.36	90.13	99.33	109.14	118.50	124.34	129.56	135.81	140.17
z_P	−2.576	−2.326	−1.960	−1.645	−1.282	−0.6745	0.0000	+0.6745	+1.282	+1.645	+1.960	+2.326	+2.576

* Use explained in Sec. 3.2.2, 4.2, 4.3, 4.6, and 4.7.1. For a number of degrees of freedom $f > 100$, take

$$\chi_f^2(P) = f\left(1 - \frac{2}{9f} + z_P \sqrt{\frac{2}{9f}}\right)^3$$

where z_P is negative for $P < 0.5$. This table was adapted, with the permission of the author and the editor, from C. M. Thompson, BIOMETRIKA, Vol. 32 (1941–42), pp. 188–89.

TABLE 5. CRITICAL VALUES FOR THE F DISTRIBUTION *

$$P(F) = \int_F^\infty \frac{\left(\dfrac{f_1+f_2-2}{2}\right)!}{\left(\dfrac{f_1-2}{2}\right)!\left(\dfrac{f_2-2}{2}\right)!} f_1^{f_1/2} f_2^{f_2/2} F^{(f_1-2)/2} (f_2 + f_1 F)^{-(f_1+f_2)/2}\, dF$$

The number of degrees of freedom for the numerator is f_1, for the denominator, f_2.

$$P(F) = 0.10$$

f_2 \ f_1	1	2	3	4	5	6	7	8	9	10	12	15	20	24	30	40	60	120	∞
1	39.86	49.50	53.59	55.83	57.24	58.20	58.91	59.44	59.86	60.20	60.70	61.22	61.74	62.00	62.26	62.53	62.79	63.06	63.33
2	8.53	9.00	9.16	9.24	9.29	9.33	9.35	9.37	9.38	9.39	9.41	9.42	9.44	9.45	9.46	9.47	9.47	9.48	9.49
3	5.54	5.46	5.39	5.34	5.31	5.28	5.27	5.25	5.24	5.23	5.22	5.20	5.18	5.18	5.17	5.16	5.15	5.14	5.13
4	4.54	4.32	4.19	4.11	4.05	4.01	3.98	3.95	3.94	3.92	3.90	3.87	3.84	3.83	3.82	3.80	3.79	3.78	3.76
5	4.06	3.78	3.62	3.52	3.45	3.40	3.37	3.34	3.32	3.30	3.27	3.24	3.21	3.19	3.17	3.16	3.14	3.12	3.10
6	3.78	3.46	3.29	3.18	3.11	3.05	3.01	2.98	2.96	2.94	2.90	2.87	2.84	2.82	2.80	2.78	2.76	2.74	2.72
7	3.59	3.26	3.07	2.96	2.88	2.83	2.78	2.75	2.72	2.70	2.67	2.63	2.59	2.58	2.56	2.54	2.51	2.49	2.47
8	3.46	3.11	2.92	2.81	2.73	2.67	2.62	2.59	2.56	2.54	2.50	2.46	2.42	2.40	2.38	2.36	2.34	2.32	2.29
9	3.36	3.01	2.81	2.69	2.61	2.55	2.51	2.47	2.44	2.42	2.38	2.34	2.30	2.28	2.25	2.23	2.21	2.18	2.16
10	3.28	2.92	2.73	2.61	2.52	2.46	2.41	2.38	2.35	2.32	2.28	2.24	2.20	2.18	2.16	2.13	2.11	2.08	2.06

f_1 / f_2	1	2	3	4	5	6	7	8	9	10	12	15	20	24	30	40	60	120	∞
11	3.23	2.86	2.66	2.54	2.45	2.39	2.34	2.30	2.27	2.25	2.21	2.17	2.12	2.10	2.08	2.05	2.03	2.00	1.97
12	3.18	2.81	2.61	2.48	2.39	2.33	2.28	2.24	2.21	2.19	2.15	2.10	2.06	2.04	2.01	1.99	1.96	1.93	1.90
13	3.14	2.76	2.56	2.43	2.35	2.28	2.23	2.20	2.16	2.14	2.10	2.05	2.01	1.98	1.96	1.93	1.90	1.88	1.85
14	3.10	2.73	2.52	2.39	2.31	2.24	2.19	2.15	2.12	2.10	2.05	2.01	1.96	1.94	1.91	1.89	1.86	1.83	1.80
15	3.07	2.70	2.49	2.36	2.27	2.21	2.16	2.12	2.09	2.06	2.02	1.97	1.92	1.90	1.87	1.85	1.82	1.79	1.76
16	3.05	2.67	2.46	2.33	2.24	2.18	2.13	2.09	2.06	2.03	1.99	1.94	1.89	1.87	1.84	1.81	1.78	1.75	1.72
17	3.03	2.64	2.44	2.31	2.22	2.15	2.10	2.06	2.03	2.00	1.96	1.91	1.86	1.84	1.81	1.78	1.75	1.72	1.69
18	3.01	2.62	2.42	2.29	2.20	2.13	2.08	2.04	2.00	1.98	1.93	1.89	1.84	1.81	1.78	1.75	1.72	1.69	1.66
19	2.99	2.61	2.40	2.27	2.18	2.11	2.06	2.02	1.98	1.96	1.91	1.86	1.81	1.79	1.76	1.73	1.70	1.67	1.63
20	2.97	2.59	2.38	2.25	2.16	2.09	2.04	2.00	1.96	1.94	1.89	1.84	1.79	1.77	1.74	1.71	1.68	1.64	1.61
21	2.96	2.57	2.36	2.23	2.14	2.08	2.02	1.98	1.95	1.92	1.88	1.83	1.78	1.75	1.72	1.69	1.66	1.62	1.59
22	2.95	2.56	2.35	2.22	2.13	2.06	2.01	1.97	1.93	1.90	1.86	1.81	1.76	1.73	1.70	1.67	1.64	1.60	1.57
23	2.94	2.55	2.34	2.21	2.11	2.05	1.99	1.95	1.92	1.89	1.84	1.80	1.74	1.72	1.69	1.66	1.62	1.59	1.55
24	2.93	2.54	2.33	2.19	2.10	2.04	1.98	1.94	1.91	1.88	1.83	1.78	1.73	1.70	1.67	1.64	1.61	1.57	1.53
25	2.92	2.53	2.32	2.18	2.09	2.02	1.97	1.93	1.89	1.87	1.82	1.77	1.72	1.69	1.66	1.63	1.59	1.56	1.52
26	2.91	2.52	2.31	2.17	2.08	2.01	1.96	1.92	1.88	1.86	1.81	1.76	1.71	1.68	1.65	1.61	1.58	1.54	1.50
27	2.90	2.51	2.30	2.17	2.07	2.00	1.95	1.91	1.87	1.85	1.80	1.75	1.70	1.67	1.64	1.60	1.57	1.53	1.49
28	2.89	2.50	2.29	2.16	2.06	2.00	1.94	1.90	1.87	1.84	1.79	1.74	1.69	1.66	1.63	1.59	1.56	1.52	1.48
29	2.89	2.50	2.28	2.15	2.06	1.99	1.93	1.89	1.86	1.83	1.78	1.73	1.68	1.65	1.62	1.58	1.55	1.51	1.47
30	2.88	2.49	2.28	2.14	2.05	1.98	1.93	1.88	1.85	1.82	1.77	1.72	1.67	1.64	1.61	1.57	1.54	1.50	1.46
40	2.84	2.44	2.23	2.09	2.00	1.93	1.87	1.83	1.79	1.76	1.71	1.66	1.61	1.57	1.54	1.51	1.47	1.42	1.38
60	2.79	2.39	2.18	2.04	1.95	1.87	1.82	1.77	1.74	1.71	1.66	1.60	1.54	1.51	1.48	1.44	1.40	1.35	1.29
120	2.75	2.35	2.13	1.99	1.90	1.82	1.77	1.72	1.68	1.65	1.60	1.54	1.48	1.45	1.41	1.37	1.32	1.26	1.19
∞	2.71	2.30	2.08	1.94	1.85	1.77	1.72	1.67	1.63	1.60	1.55	1.49	1.42	1.38	1.34	1.30	1.24	1.17	1.00

* Use explained in Sec. 3.3.2 and 5.4–5.6. The relation $F_\alpha(f_1, f_2) = 1/F_{1-\alpha}(f_2, f_1)$ holds. Critical values for $P > 0.5$ are obtained from $F_P(f_1, f_2) = 1/F_{1-P}(f_2, f_1)$. The reciprocals of the numbers of degrees of freedom, rather than the numbers themselves, should be used for linear interpolation. For example, for $f_1 = 18$, $f_2 = 240$, and $P = 2.5\%$

$$F = 1.765 + \frac{(1/18) - (1/20)}{(1/15) - (1/20)} (0.12) = 1.80$$

This table was adapted, with the permission of the authors and the editor, from M. Merrington and C. M. Thompson, "Tables of Percentage Points of the Inverted Beta (F) Distribution," BIOMETRIKA, Vol. 33 (1943–46), pp. 73–88. The BIOMETRIKA tables give four decimal places.

TABLE 5. (Contd.)

$$P(F) = 0.05$$

f_2 \ f_1	1	2	3	4	5	6	7	8	9	10	12	15	20	24	30	40	60	120	∞
1	161.45	199.50	215.71	224.58	230.16	233.99	236.77	238.88	240.54	241.88	243.91	245.95	248.01	249.05	250.09	251.14	252.20	253.25	254.32
2	18.51	19.00	19.16	19.25	19.30	19.33	19.35	19.37	19.38	19.40	19.41	19.43	19.45	19.45	19.46	19.47	19.48	19.49	19.50
3	10.13	9.55	9.28	9.12	9.01	8.94	8.89	8.85	8.81	8.79	8.74	8.70	8.66	8.64	8.62	8.59	8.57	8.55	8.53
4	7.71	6.94	6.59	6.39	6.26	6.16	6.09	6.04	6.00	5.96	5.91	5.86	5.80	5.77	5.75	5.72	5.69	5.66	5.63
5	6.61	5.79	5.41	5.19	5.05	4.95	4.88	4.82	4.77	4.74	4.68	4.62	4.56	4.53	4.50	4.46	4.43	4.40	4.36
6	5.99	5.14	4.76	4.53	4.39	4.28	4.21	4.15	4.10	4.06	4.00	3.94	3.87	3.84	3.81	3.77	3.74	3.70	3.67
7	5.59	4.74	4.35	4.12	3.97	3.87	3.79	3.73	3.68	3.64	3.57	3.51	3.44	3.41	3.38	3.34	3.30	3.27	3.23
8	5.32	4.46	4.07	3.84	3.69	3.58	3.50	3.44	3.39	3.35	3.28	3.22	3.15	3.12	3.08	3.04	3.01	2.97	2.93
9	5.12	4.26	3.86	3.63	3.48	3.37	3.29	3.23	3.18	3.14	3.07	3.01	2.94	2.90	2.86	2.83	2.79	2.75	2.71
10	4.96	4.10	3.71	3.48	3.33	3.22	3.14	3.07	3.02	2.98	2.91	2.84	2.77	2.74	2.70	2.66	2.62	2.58	2.54
11	4.84	3.98	3.59	3.36	3.20	3.09	3.01	2.95	2.90	2.85	2.79	2.72	2.65	2.61	2.57	2.53	2.49	2.45	2.40
12	4.75	3.89	3.49	3.26	3.11	3.00	2.91	2.85	2.80	2.75	2.69	2.62	2.54	2.51	2.47	2.43	2.38	2.34	2.30
13	4.67	3.81	3.41	3.18	3.03	2.92	2.83	2.77	2.71	2.67	2.60	2.53	2.46	2.42	2.38	2.34	2.30	2.25	2.21
14	4.60	3.74	3.34	3.11	2.96	2.85	2.76	2.70	2.65	2.60	2.53	2.46	2.39	2.35	2.31	2.27	2.22	2.18	2.13
15	4.54	3.68	3.29	3.06	2.90	2.79	2.71	2.64	2.59	2.54	2.48	2.40	2.33	2.29	2.25	2.20	2.16	2.11	2.07
16	4.49	3.63	3.24	3.01	2.85	2.74	2.66	2.59	2.54	2.49	2.42	2.35	2.28	2.24	2.19	2.15	2.11	2.06	2.01
17	4.45	3.59	3.20	2.96	2.81	2.70	2.61	2.55	2.49	2.45	2.38	2.31	2.23	2.19	2.15	2.10	2.06	2.01	1.96
18	4.41	3.55	3.16	2.93	2.77	2.66	2.58	2.51	2.46	2.41	2.34	2.27	2.19	2.15	2.11	2.06	2.02	1.97	1.92
19	4.38	3.52	3.13	2.90	2.74	2.63	2.54	2.48	2.42	2.38	2.31	2.23	2.16	2.11	2.07	2.03	1.98	1.93	1.88
20	4.35	3.49	3.10	2.87	2.71	2.60	2.51	2.45	2.39	2.35	2.28	2.20	2.12	2.08	2.04	1.99	1.95	1.90	1.84
21	4.32	3.47	3.07	2.84	2.68	2.57	2.49	2.42	2.37	2.32	2.25	2.18	2.10	2.05	2.01	1.96	1.92	1.87	1.81
22	4.30	3.44	3.05	2.82	2.66	2.55	2.46	2.40	2.34	2.30	2.23	2.15	2.07	2.03	1.98	1.94	1.89	1.84	1.78
23	4.28	3.42	3.03	2.80	2.64	2.53	2.44	2.37	2.32	2.27	2.20	2.13	2.05	2.01	1.96	1.91	1.86	1.81	1.76
24	4.26	3.40	3.01	2.78	2.62	2.51	2.42	2.36	2.30	2.25	2.18	2.11	2.03	1.98	1.94	1.89	1.84	1.79	1.73
25	4.24	3.39	2.99	2.76	2.60	2.49	2.40	2.34	2.28	2.24	2.16	2.09	2.01	1.96	1.92	1.87	1.82	1.77	1.71
26	4.23	3.37	2.98	2.74	2.59	2.47	2.39	2.32	2.27	2.22	2.15	2.07	1.99	1.95	1.90	1.85	1.80	1.75	1.69
27	4.21	3.35	2.96	2.73	2.57	2.46	2.37	2.31	2.25	2.20	2.13	2.06	1.97	1.93	1.88	1.84	1.79	1.73	1.67
28	4.20	3.34	2.95	2.71	2.56	2.45	2.36	2.29	2.24	2.19	2.12	2.04	1.96	1.91	1.87	1.82	1.77	1.71	1.65
29	4.18	3.33	2.93	2.70	2.55	2.43	2.35	2.28	2.22	2.18	2.10	2.03	1.94	1.90	1.85	1.81	1.75	1.70	1.64
30	4.17	3.32	2.92	2.69	2.53	2.42	2.33	2.27	2.21	2.16	2.09	2.01	1.93	1.89	1.84	1.79	1.74	1.68	1.62
40	4.08	3.23	2.84	2.61	2.45	2.34	2.25	2.18	2.12	2.08	2.00	1.92	1.84	1.79	1.74	1.69	1.64	1.58	1.51
60	4.00	3.15	2.76	2.53	2.37	2.25	2.17	2.10	2.04	1.99	1.92	1.84	1.75	1.70	1.65	1.59	1.53	1.47	1.39
120	3.92	3.07	2.68	2.45	2.29	2.18	2.09	2.02	1.96	1.91	1.83	1.75	1.66	1.61	1.55	1.50	1.43	1.35	1.25
∞	3.84	3.00	2.60	2.37	2.21	2.10	2.01	1.94	1.88	1.83	1.75	1.67	1.57	1.52	1.46	1.39	1.32	1.22	1.00

$P(F) = 0.025$

f_2 \ f_1	1	2	3	4	5	6	7	8	9	10	12	15	20	24	30	40	60	120	∞
1	647.79	799.50	864.16	899.58	921.85	937.11	948.22	956.66	963.28	968.63	976.71	984.87	993.10	997.25	1001.4	1005.6	1009.8	1014.0	1018.3
2	38.51	39.00	39.16	39.25	39.30	39.33	39.36	39.37	39.39	39.40	39.42	39.43	39.45	39.46	39.46	39.47	39.48	39.49	39.50
3	17.44	16.04	15.44	15.10	14.88	14.74	14.62	14.54	14.47	14.42	14.34	14.25	14.17	14.12	14.08	14.04	13.99	13.95	13.90
4	12.22	10.65	9.98	9.60	9.36	9.20	9.07	8.98	8.90	8.84	8.75	8.66	8.56	8.51	8.46	8.41	8.36	8.31	8.26
5	10.01	8.43	7.76	7.39	7.15	6.98	6.85	6.76	6.68	6.62	6.52	6.43	6.33	6.28	6.23	6.18	6.12	6.07	6.02
6	8.81	7.26	6.60	6.23	5.99	5.82	5.70	5.60	5.52	5.46	5.37	5.27	5.17	5.12	5.07	5.01	4.96	4.90	4.85
7	8.07	6.54	5.89	5.52	5.29	5.12	4.99	4.90	4.82	4.76	4.67	4.57	4.47	4.42	4.36	4.31	4.25	4.20	4.14
8	7.57	6.06	5.42	5.05	4.82	4.65	4.53	4.43	4.36	4.30	4.20	4.10	4.00	3.95	3.89	3.84	3.78	3.73	3.67
9	7.21	5.71	5.08	4.72	4.48	4.32	4.20	4.10	4.03	3.96	3.87	3.77	3.67	3.61	3.56	3.51	3.45	3.39	3.33
10	6.94	5.46	4.83	4.47	4.24	4.07	3.95	3.85	3.78	3.72	3.62	3.52	3.42	3.37	3.31	3.26	3.20	3.14	3.08
11	6.72	5.26	4.63	4.28	4.04	3.88	3.76	3.66	3.59	3.53	3.43	3.33	3.23	3.17	3.12	3.06	3.00	2.94	2.88
12	6.55	5.10	4.47	4.12	3.89	3.73	3.61	3.51	3.44	3.37	3.28	3.18	3.07	3.02	2.96	2.91	2.85	2.79	2.72
13	6.41	4.97	4.35	4.00	3.77	3.60	3.48	3.39	3.31	3.25	3.15	3.05	2.95	2.89	2.84	2.78	2.72	2.66	2.60
14	6.30	4.86	4.24	3.89	3.66	3.50	3.38	3.29	3.21	3.15	3.05	2.95	2.84	2.79	2.73	2.67	2.61	2.55	2.49
15	6.20	4.76	4.15	3.80	3.58	3.41	3.29	3.20	3.12	3.06	2.96	2.86	2.76	2.70	2.64	2.58	2.52	2.46	2.40
16	6.12	4.69	4.08	3.73	3.50	3.34	3.22	3.12	3.05	2.99	2.89	2.79	2.68	2.63	2.57	2.51	2.45	2.38	2.32
17	6.04	4.62	4.01	3.66	3.44	3.28	3.16	3.06	2.98	2.92	2.82	2.72	2.62	2.56	2.50	2.44	2.38	2.32	2.25
18	5.98	4.56	3.95	3.61	3.38	3.22	3.10	3.01	2.93	2.87	2.77	2.67	2.56	2.50	2.44	2.38	2.32	2.26	2.19
19	5.92	4.51	3.90	3.56	3.33	3.17	3.05	2.96	2.88	2.82	2.72	2.62	2.51	2.45	2.39	2.33	2.27	2.20	2.13
20	5.87	4.46	3.86	3.51	3.29	3.13	3.01	2.91	2.84	2.77	2.68	2.57	2.46	2.41	2.35	2.29	2.22	2.16	2.09
21	5.83	4.42	3.82	3.48	3.25	3.09	2.97	2.87	2.80	2.73	2.64	2.53	2.42	2.37	2.31	2.25	2.18	2.11	2.04
22	5.79	4.38	3.78	3.44	3.22	3.05	2.93	2.84	2.76	2.70	2.60	2.50	2.39	2.33	2.27	2.21	2.14	2.08	2.00
23	5.75	4.35	3.75	3.41	3.18	3.02	2.90	2.81	2.73	2.67	2.57	2.47	2.36	2.30	2.24	2.18	2.11	2.04	1.97
24	5.72	4.32	3.72	3.38	3.15	2.99	2.87	2.78	2.70	2.64	2.54	2.44	2.33	2.27	2.21	2.15	2.08	2.01	1.94
25	5.69	4.29	3.69	3.35	3.13	2.97	2.85	2.75	2.68	2.61	2.51	2.41	2.30	2.24	2.18	2.12	2.05	1.98	1.91
26	5.66	4.27	3.67	3.33	3.10	2.94	2.82	2.73	2.65	2.59	2.49	2.39	2.28	2.22	2.16	2.09	2.03	1.95	1.88
27	5.63	4.24	3.65	3.31	3.08	2.92	2.80	2.71	2.63	2.57	2.47	2.36	2.25	2.19	2.13	2.07	2.00	1.93	1.85
28	5.61	4.22	3.63	3.29	3.06	2.90	2.78	2.69	2.61	2.55	2.45	2.34	2.23	2.17	2.11	2.05	1.98	1.91	1.83
29	5.59	4.20	3.61	3.27	3.04	2.88	2.76	2.67	2.59	2.53	2.43	2.32	2.21	2.15	2.09	2.03	1.96	1.89	1.81
30	5.57	4.18	3.59	3.25	3.03	2.87	2.75	2.65	2.57	2.51	2.41	2.31	2.20	2.14	2.07	2.01	1.94	1.87	1.79
40	5.42	4.05	3.46	3.13	2.90	2.74	2.62	2.53	2.45	2.39	2.29	2.18	2.07	2.01	1.94	1.88	1.80	1.72	1.64
60	5.29	3.93	3.34	3.01	2.79	2.63	2.51	2.41	2.33	2.27	2.17	2.06	1.94	1.88	1.82	1.74	1.67	1.58	1.48
120	5.15	3.80	3.23	2.89	2.67	2.52	2.39	2.30	2.22	2.16	2.05	1.94	1.82	1.76	1.69	1.61	1.53	1.43	1.31
∞	5.02	3.69	3.12	2.79	2.57	2.41	2.29	2.19	2.11	2.05	1.94	1.83	1.71	1.64	1.57	1.48	1.39	1.27	1.00

See notes at the beginning of the table.

TABLE 5. (Contd.)

$P(F) = 0.01$

f_2 \ f_1	1	2	3	4	5	6	7	8	9	10	12	15	20	24	30	40	60	120	∞
1	4052.2	4999.5	5403.3	5624.6	5763.7	5859.0	5928.3	5981.6	6022.5	6055.8	6106.3	6157.3	6208.7	6234.6	6260.7	6286.8	6313.0	6339.4	6366.0
2	98.50	99.00	99.17	99.25	99.30	99.33	99.36	99.37	99.39	99.40	99.42	99.43	99.45	99.46	99.47	99.47	99.48	99.49	99.50
3	34.12	30.82	29.46	28.71	28.24	27.91	27.67	27.49	27.34	27.23	27.05	26.87	26.69	26.60	26.50	26.41	26.32	26.22	26.12
4	21.20	18.00	16.69	15.98	15.52	15.21	14.98	14.80	14.66	14.55	14.37	14.20	14.02	13.93	13.84	13.74	13.65	13.56	13.46
5	16.26	13.27	12.06	11.39	10.97	10.67	10.46	10.29	10.16	10.05	9.89	9.72	9.55	9.47	9.38	9.29	9.20	9.11	9.02
6	13.74	10.92	9.78	9.15	8.75	8.47	8.26	8.10	7.98	7.87	7.72	7.56	7.40	7.31	7.23	7.14	7.06	6.97	6.88
7	12.25	9.55	8.45	7.85	7.46	7.19	6.99	6.84	6.72	6.62	6.47	6.31	6.16	6.07	5.99	5.91	5.82	5.74	5.65
8	11.26	8.65	7.59	7.01	6.63	6.37	6.18	6.03	5.91	5.81	5.67	5.52	5.36	5.28	5.20	5.12	5.03	4.95	4.86
9	10.56	8.02	6.99	6.42	6.06	5.80	5.61	5.47	5.35	5.26	5.11	4.96	4.81	4.73	4.65	4.57	4.48	4.40	4.31
10	10.04	7.56	6.55	5.99	5.64	5.39	5.20	5.06	4.94	4.85	4.71	4.56	4.41	4.33	4.25	4.17	4.08	4.00	3.91
11	9.65	7.21	6.22	5.67	5.32	5.07	4.89	4.74	4.63	4.54	4.40	4.25	4.10	4.02	3.94	3.86	3.78	3.69	3.60
12	9.33	6.93	5.95	5.41	5.06	4.82	4.64	4.50	4.39	4.30	4.16	4.01	3.86	3.78	3.70	3.62	3.54	3.45	3.36
13	9.07	6.70	5.74	5.21	4.86	4.62	4.44	4.30	4.19	4.10	3.96	3.82	3.66	3.59	3.51	3.43	3.34	3.25	3.17
14	8.86	6.51	5.56	5.04	4.70	4.46	4.28	4.14	4.03	3.94	3.80	3.66	3.51	3.43	3.35	3.27	3.18	3.09	3.00
15	8.68	6.36	5.42	4.89	4.56	4.32	4.14	4.00	3.89	3.80	3.67	3.52	3.37	3.29	3.21	3.13	3.05	2.96	2.87
16	8.53	6.23	5.29	4.77	4.44	4.20	4.03	3.89	3.78	3.69	3.55	3.41	3.26	3.18	3.10	3.02	2.93	2.84	2.75
17	8.40	6.11	5.18	4.67	4.34	4.10	3.93	3.79	3.68	3.59	3.46	3.31	3.16	3.08	3.00	2.92	2.83	2.75	2.65
18	8.29	6.01	5.09	4.58	4.25	4.01	3.84	3.71	3.60	3.51	3.37	3.23	3.08	3.00	2.92	2.84	2.75	2.66	2.57
19	8.18	5.93	5.01	4.50	4.17	3.94	3.77	3.63	3.52	3.43	3.30	3.15	3.00	2.92	2.84	2.76	2.67	2.58	2.49
20	8.10	5.85	4.94	4.43	4.10	3.87	3.70	3.56	3.46	3.37	3.23	3.09	2.94	2.86	2.78	2.69	2.61	2.52	2.42
21	8.02	5.78	4.87	4.37	4.04	3.81	3.64	3.51	3.40	3.31	3.17	3.03	2.88	2.80	2.72	2.64	2.55	2.46	2.36
22	7.95	5.72	4.82	4.31	3.99	3.76	3.59	3.45	3.35	3.26	3.12	2.98	2.83	2.75	2.67	2.58	2.50	2.40	2.31
23	7.88	5.66	4.76	4.26	3.94	3.71	3.54	3.41	3.30	3.21	3.07	2.93	2.78	2.70	2.62	2.54	2.45	2.35	2.26
24	7.82	5.61	4.72	4.22	3.90	3.67	3.50	3.36	3.26	3.17	3.03	2.89	2.74	2.66	2.58	2.49	2.40	2.31	2.21
25	7.77	5.57	4.68	4.18	3.86	3.63	3.46	3.32	3.22	3.13	2.99	2.85	2.70	2.62	2.54	2.45	2.36	2.27	2.17
26	7.72	5.53	4.64	4.14	3.82	3.59	3.42	3.29	3.18	3.09	2.96	2.81	2.66	2.58	2.50	2.42	2.33	2.23	2.13
27	7.68	5.49	4.60	4.11	3.78	3.56	3.39	3.26	3.15	3.06	2.93	2.78	2.63	2.55	2.47	2.38	2.29	2.20	2.10
28	7.64	5.45	4.57	4.07	3.75	3.53	3.36	3.23	3.12	3.03	2.90	2.75	2.60	2.52	2.44	2.35	2.26	2.17	2.06
29	7.60	5.42	4.54	4.04	3.73	3.50	3.33	3.20	3.09	3.00	2.87	2.73	2.57	2.49	2.41	2.33	2.23	2.14	2.03
30	7.56	5.39	4.51	4.02	3.70	3.47	3.30	3.17	3.07	2.98	2.84	2.70	2.55	2.47	2.39	2.30	2.21	2.11	2.01
40	7.31	5.18	4.31	3.83	3.51	3.29	3.12	2.99	2.89	2.80	2.66	2.52	2.37	2.29	2.20	2.11	2.02	1.92	1.80
60	7.08	4.98	4.13	3.65	3.34	3.12	2.95	2.82	2.72	2.63	2.50	2.35	2.20	2.12	2.03	1.94	1.84	1.73	1.60
120	6.85	4.79	3.95	3.48	3.17	2.96	2.79	2.66	2.56	2.47	2.34	2.19	2.03	1.95	1.86	1.76	1.66	1.53	1.38
∞	6.63	4.61	3.78	3.32	3.02	2.80	2.64	2.51	2.41	2.32	2.18	2.04	1.88	1.79	1.70	1.59	1.47	1.32	1.00

$P(F) = 0.005$

f_2 \ f_1	∞	120	60	40	30	24	20	15	12	10	9	8	7	6	5	4	3	2	1
1	25465	25359	25253	25148	25044	24940	24836	24630	24426	24224	24091	23925	23715	23437	23056	22500	21615	20000	16211
2	199.51	199.49	199.48	199.47	199.47	199.46	199.45	199.43	199.42	199.40	199.39	199.37	199.36	199.33	199.30	199.25	199.17	199.00	198.50
3	41.83	41.99	42.15	42.31	42.47	42.62	42.78	43.08	43.39	43.69	43.88	44.13	44.43	44.84	45.39	46.20	47.47	49.80	55.55
4	19.32	19.47	19.61	19.75	19.89	20.03	20.17	20.44	20.70	20.97	21.14	21.35	21.62	21.98	22.46	23.16	24.26	26.28	31.33
5	12.14	12.27	12.40	12.53	12.66	12.78	12.90	13.15	13.38	13.62	13.77	13.96	14.20	14.51	14.94	15.56	16.53	18.31	22.78
6	8.88	9.00	9.12	9.24	9.36	9.47	9.59	9.81	10.03	10.25	10.39	10.57	10.79	11.07	11.46	12.03	12.92	14.54	18.64
7	7.08	7.19	7.31	7.42	7.53	7.64	7.75	7.97	8.18	8.38	8.51	8.68	8.89	9.16	9.52	10.05	10.88	12.40	16.24
8	5.95	6.06	6.18	6.29	6.40	6.50	6.61	6.81	7.01	7.21	7.34	7.50	7.69	7.95	8.30	8.81	9.60	11.04	14.69
9	5.19	5.30	5.41	5.52	5.62	5.73	5.83	6.03	6.23	6.42	6.54	6.69	6.88	7.13	7.47	7.96	8.72	10.11	13.61
10	4.64	4.75	4.86	4.97	5.07	5.17	5.27	5.47	5.66	5.85	5.97	6.12	6.30	6.54	6.87	7.34	8.08	9.43	12.83
11	4.23	4.34	4.44	4.55	4.65	4.76	4.86	5.05	5.24	5.42	5.54	5.68	5.86	6.10	6.42	6.88	7.60	8.91	12.23
12	3.90	4.01	4.12	4.23	4.33	4.43	4.53	4.72	4.91	5.09	5.20	5.35	5.52	5.76	6.07	6.52	7.23	8.51	11.75
13	3.65	3.76	3.87	3.97	4.07	4.17	4.27	4.46	4.64	4.82	4.94	5.08	5.25	5.48	5.79	6.23	6.93	8.19	11.37
14	3.44	3.55	3.66	3.76	3.86	3.96	4.06	4.25	4.43	4.60	4.72	4.86	5.03	5.26	5.56	6.00	6.68	7.92	11.06
15	3.26	3.37	3.48	3.58	3.69	3.79	3.88	4.07	4.25	4.42	4.54	4.67	4.85	5.07	5.37	5.80	6.48	7.70	10.80
16	3.11	3.22	3.33	3.44	3.54	3.64	3.73	3.92	4.10	4.27	4.38	4.52	4.69	4.91	5.21	5.64	6.30	7.51	10.58
17	2.98	3.10	3.21	3.31	3.41	3.51	3.61	3.79	3.97	4.14	4.25	4.39	4.56	4.78	5.07	5.50	6.16	7.35	10.38
18	2.87	2.99	3.10	3.20	3.30	3.40	3.50	3.68	3.86	4.03	4.14	4.28	4.44	4.66	4.96	5.37	6.03	7.21	10.22
19	2.78	2.89	3.00	3.11	3.21	3.31	3.40	3.59	3.76	3.93	4.04	4.18	4.34	4.56	4.85	5.27	5.92	7.09	10.07
20	2.69	2.81	2.92	3.02	3.12	3.22	3.32	3.50	3.68	3.85	3.96	4.09	4.26	4.47	4.76	5.17	5.82	6.99	9.94
21	2.61	2.73	2.84	2.95	3.05	3.15	3.24	3.43	3.60	3.77	3.88	4.01	4.18	4.39	4.68	5.09	5.73	6.89	9.83
22	2.55	2.66	2.77	2.88	2.98	3.08	3.18	3.36	3.54	3.70	3.81	3.94	4.11	4.32	4.61	5.02	5.65	6.81	9.73
23	2.48	2.60	2.71	2.82	2.92	3.02	3.12	3.30	3.47	3.64	3.75	3.88	4.05	4.26	4.54	4.95	5.58	6.73	9.63
24	2.43	2.55	2.66	2.77	2.87	2.97	3.06	3.25	3.42	3.59	3.69	3.83	3.99	4.20	4.49	4.89	5.52	6.66	9.55
25	2.38	2.50	2.61	2.72	2.82	2.92	3.01	3.20	3.37	3.54	3.64	3.78	3.94	4.15	4.43	4.84	5.46	6.60	9.48
26	2.33	2.45	2.56	2.67	2.77	2.87	2.97	3.15	3.33	3.49	3.60	3.73	3.89	4.10	4.38	4.79	5.41	6.54	9.41
27	2.29	2.41	2.52	2.63	2.73	2.83	2.93	3.11	3.28	3.45	3.56	3.69	3.85	4.06	4.34	4.74	5.36	6.49	9.34
28	2.25	2.37	2.48	2.59	2.69	2.79	2.89	3.07	3.25	3.41	3.52	3.65	3.81	4.02	4.30	4.70	5.32	6.44	9.28
29	2.21	2.33	2.45	2.56	2.66	2.76	2.86	3.04	3.21	3.38	3.48	3.61	3.77	3.98	4.26	4.66	5.28	6.40	9.23
30	2.18	2.30	2.42	2.52	2.63	2.73	2.82	3.01	3.18	3.34	3.45	3.58	3.74	3.95	4.23	4.62	5.24	6.35	9.18
40	1.93	2.06	2.18	2.30	2.40	2.50	2.60	2.78	2.95	3.12	3.22	3.35	3.51	3.71	3.99	4.37	4.98	6.07	8.83
60	1.69	1.83	1.96	2.08	2.19	2.29	2.39	2.57	2.74	2.90	3.01	3.13	3.29	3.49	3.76	4.14	4.73	5.80	8.49
120	1.43	1.61	1.75	1.87	1.98	2.09	2.19	2.37	2.54	2.71	2.81	2.93	3.09	3.28	3.55	3.92	4.50	5.54	8.18
∞	1.00	1.36	1.53	1.67	1.79	1.90	2.00	2.19	2.36	2.52	2.62	2.74	2.90	3.09	3.35	3.72	4.28	5.30	7.88

See notes at the beginning of the table.

TABLE 6. CRITICAL VALUES FOR THE M DISTRIBUTION AT THE 5% SIGNIFICANCE LEVEL *

c_1 / k	0.0	0.5	1.0	1.5	2.0	2.5	3.0	3.5	4.0	4.5	5.0	6.0	7.0	8.0	9.0	10.0	12.0	14.0
3 (a)	5.99	6.47	6.89	7.20	7.38	7.39	7.22											
3 (b)	5.99	6.22	6.43	6.64	6.84	7.03	7.22											
4 (a)	7.81	8.24	8.63	8.96	9.21	9.38	9.43	9.37	9.18									
4 (b)	7.81	8.00	8.17	8.35	8.52	8.69	8.85	9.02	9.18									
5 (a)	9.49	9.88	10.24	10.57	10.86	11.08	11.24	11.32	11.31	11.21	11.02							
5 (b)	9.49	9.65	9.80	9.96	10.11	10.27	10.42	10.57	10.72	10.87	11.02							
6 (a)	11.07	11.43	11.78	12.11	12.40	12.65	12.86	13.01	13.11	13.14	13.10	12.78						
6 (b)	11.07	11.22	11.36	11.51	11.65	11.79	11.94	12.08	12.22	12.36	12.50	12.78						
7 (a)	12.59	12.94	13.27	13.59	13.88	14.15	14.38	14.58	14.73	14.83	14.88	14.81	14.49					
7 (b)	12.59	12.73	12.87	13.00	13.14	13.27	13.41	13.55	13.68	13.82	13.95	14.22	14.49					
8 (a)	14.07	14.40	14.72	15.03	15.32	15.60	15.84	16.06	16.25	16.40	16.51	16.60	16.49	16.16				
8 (b)	14.07	14.20	14.33	14.46	14.59	14.72	14.85	14.98	15.11	15.25	15.38	15.64	15.90	16.16				
9 (a)	15.51	15.83	16.14	16.44	16.73	17.01	17.26	17.49	17.70	17.88	18.03	18.22	18.26	18.12	17.79			
9 (b)	15.51	15.63	15.76	15.89	16.02	16.14	16.27	16.40	16.52	16.65	16.78	17.03	17.29	17.54	17.79			
10 (a)	16.92	17.23	17.54	17.83	18.12	18.39	18.65	18.89	19.11	19.31	19.48	19.75	19.89	19.89	19.73	19.40		
10 (b)	16.92	17.04	17.17	17.29	17.41	17.54	17.66	17.79	17.91	18.04	18.16	18.41	18.66	18.91	19.16	19.40		
11 (a)	18.31	18.61	18.91	19.20	19.48	19.76	20.02	20.26	20.49	20.70	20.89	21.21	21.42	21.52	21.49	21.32		
11 (b)	18.31	18.43	18.55	18.67	18.79	18.91	19.04	19.16	19.28	19.40	19.52	19.77	20.01	20.26	20.50	20.75		
12 (a)	19.68	19.97	20.26	20.55	20.83	21.10	21.36	21.61	21.84	22.06	22.27	22.62	22.88	23.06	23.12	23.07	22.56	
12 (b)	19.68	19.79	19.91	20.03	20.15	20.27	20.39	20.51	20.63	20.75	20.87	21.12	21.36	21.60	21.84	22.08	22.56	
13 (a)	21.03	21.32	21.60	21.89	22.16	22.43	22.69	22.94	23.18	23.40	23.62	23.99	24.30	24.53	24.66	24.70	24.44	
13 (b)	21.03	21.14	21.26	21.38	21.50	21.62	21.74	21.85	21.97	22.09	22.21	22.45	22.69	22.92	23.16	23.40	23.88	
14 (a)	22.36	22.65	22.93	23.21	23.48	23.75	24.01	24.26	24.50	24.73	24.95	25.34	25.68	25.95	26.14	26.25	26.17	25.66
14 (b)	22.36	22.48	22.60	22.71	22.83	22.95	23.06	23.18	23.30	23.42	23.53	23.77	24.00	24.24	24.48	24.71	25.19	25.66
15 (a)	23.68	23.97	24.24	24.52	24.79	25.05	25.31	25.56	25.80	26.04	26.26	26.67	27.03	27.33	27.56	27.73	27.80	27.50
15 (b)	23.68	23.80	23.92	24.03	24.15	24.26	24.38	24.50	24.61	24.73	24.85	25.08	25.31	25.55	25.78	26.01	26.48	26.95

* Use explained in Sec. 3.4.2. Reproduced with the permission of the authors and the editor from C. M. Thompson and M. Merrington, "Tables for Testing the Homogeneity of a Set of Estimated Variances," BIOMETRIKA, Vol. 33 (1943–46), pp. 296–304. These tables give 1% points also.

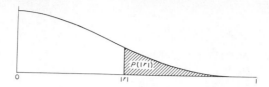

TABLE 7. Critical Absolute Values of Correlation Coefficient r *

5% points and 1% points (in boldface) for equal-tails test of hypothesis $\rho = 0$.

f	Total number of variables				f	Total number of variables			
	2	3	4	5		2	3	4	5
1	.997	.999	.999	.999	24	.388	.470	.523	.562
	1.000	**1.000**	**1.000**	**1.000**		**.496**	**.565**	**.609**	**.642**
2	.950	.975	.983	.987	25	.381	.462	.514	.553
	.990	**.995**	**.997**	**.998**		**.487**	**.555**	**.600**	**.633**
3	.878	.930	.950	.961	26	.374	.454	.506	.545
	.959	**.976**	**.983**	**.987**		**.478**	**.546**	**.590**	**.624**
4	.811	.881	.912	.930	27	.367	.446	.498	.536
	.917	**.949**	**.962**	**.970**		**.470**	**.538**	**.582**	**.615**
5	.754	.836	.874	.898	28	.361	.439	.490	.529
	.874	**.917**	**.937**	**.949**		**.463**	**.530**	**.573**	**.606**
6	.707	.795	.839	.867	29	.355	.432	.482	.521
	.834	**.886**	**.911**	**.927**		**.456**	**.522**	**.565**	**.598**
7	.666	.758	.807	.838	30	.349	.426	.476	.514
	.798	**.855**	**.885**	**.904**		**.449**	**.514**	**.558**	**.591**
8	.632	.726	.777	.811	35	.325	.397	.445	.482
	.765	**.827**	**.860**	**.882**		**.418**	**.481**	**.523**	**.556**
9	.602	.697	.750	.786	40	.304	.373	.419	.455
	.735	**.800**	**.836**	**.861**		**.393**	**.454**	**.494**	**.526**
10	.576	.671	.726	.763	45	.288	.353	.397	.432
	.708	**.776**	**.814**	**.840**		**.372**	**.430**	**.470**	**.501**
11	.553	.648	.703	.741	50	.273	.336	.379	.412
	.684	**.753**	**.793**	**.821**		**.354**	**.410**	**.449**	**.479**
12	.532	.627	.683	.722	60	.250	.308	.348	.380
	.661	**.732**	**.773**	**.802**		**.325**	**.377**	**.414**	**.442**
13	.514	.608	.664	.703	70	.232	.286	.324	.354
	.641	**.712**	**.755**	**.785**		**.302**	**.351**	**.386**	**.413**
14	.497	.590	.646	.686	80	.217	.269	.304	.332
	.623	**.694**	**.737**	**.768**		**.283**	**.330**	**.362**	**.389**
15	.482	.574	.630	.670	90	.205	.254	.288	.315
	.606	**.677**	**.721**	**.752**		**.267**	**.312**	**.343**	**.368**
16	.468	.559	.615	.655	100	.195	.241	.274	.300
	.590	**.662**	**.706**	**.738**		**.254**	**.297**	**.327**	**.351**
17	.456	.545	.601	.641	125	.174	.216	.246	.269
	.575	**.647**	**.691**	**.724**		**.228**	**.266**	**.294**	**.316**
18	.444	.532	.587	.628	150	.159	.198	.225	.247
	.561	**.633**	**.678**	**.710**		**.208**	**.244**	**.270**	**.290**
19	.433	.520	.575	.615	200	.138	.172	.196	.215
	.549	**.620**	**.665**	**.698**		**.181**	**.212**	**.234**	**.253**
20	.423	.509	.563	.604	300	.113	.141	.160	.176
	.537	**.608**	**.652**	**.685**		**.148**	**.174**	**.192**	**.208**
21	.413	.498	.552	.592	400	.098	.122	.139	.153
	.526	**.596**	**.641**	**.674**		**.128**	**.151**	**.167**	**.180**
22	.404	.488	.542	.582	500	.088	.109	.124	.137
	.515	**.585**	**.630**	**.663**		**.115**	**.135**	**.150**	**.162**
23	.396	.479	.532	.572	1000	.062	.077	.088	.097
	.505	**.574**	**.619**	**.652**		**.081**	**.096**	**.106**	**.115**

° Use explained in Sec. 6.1.4a, 6.2.5b, and 6.2.5c. The inverse square roots of the numbers of degrees of freedom, rather than the numbers themselves, should be used for linear interpolation. Reproduced, with the permission of the author and the publisher, from G. W. Snedecor, *Statistical Methods*, 4th ed., Ames, Iowa, Iowa State College Press, 1946, p. 351. Extensive tables and charts appear in F. N. David, *Tables of the Correlation Coefficient*, London, Biometrika Office, 1938.

TABLE 8. Factors for Determining Confidence Limits for Population Standard Deviation σ *

$f = n - 1$ for single sample size n

$f = n_1 + n_2 + \cdots + n_k - k$ for k sample sizes of $n_1, \cdots, n_k$

f	90%		95%		99%	
	b_1	b_2	b_1	b_2	b_1	b_2
1	.510	15.947	.446	31.910	.356	159.576
2	.578	4.415	.521	6.285	.434	14.124
3	.620	2.920	.566	3.729	.483	6.468
4	.649	2.372	.599	2.874	.519	4.396
5	.672	2.089	.624	2.453	.546	3.485
6	.690	1.915	.644	2.202	.569	2.980
7	.705	1.797	.661	2.035	.588	2.660
8	.718	1.711	.675	1.916	.604	2.439
9	.729	1.645	.688	1.826	.618	2.278
10	.739	1.593	.699	1.755	.630	2.154
11	.748	1.551	.708	1.698	.641	2.056
12	.755	1.515	.717	1.651	.651	1.976
13	.762	1.485	.725	1.611	.660	1.910
14	.769	1.460	.732	1.577	.669	1.854
15	.775	1.437	.739	1.548	.676	1.806
16	.780	1.418	.745	1.522	.683	1.764
17	.785	1.400	.750	1.499	.690	1.727
18	.790	1.384	.756	1.479	.696	1.695
19	.794	1.370	.760	1.461	.702	1.666
20	.798	1.358	.765	1.444	.707	1.640
21	.802	1.346	.769	1.429	.712	1.617
22	.805	1.335	.773	1.415	.717	1.595
23	.809	1.326	.777	1.403	.722	1.576
24	.812	1.316	.781	1.391	.726	1.558
25	.815	1.308	.784	1.380	.730	1.542
26	.818	1.300	.788	1.370	.734	1.526
27	.820	1.293	.791	1.361	.737	1.512
28	.823	1.286	.794	1.352	.741	1.499
29	.825	1.280	.796	1.344	.744	1.487
30	.828	1.274	.799	1.337	.748	1.475
40	.847	1.228	.821	1.280	.774	1.390
50	.861	1.199	.837	1.243	.793	1.337
60	.871	1.179	.849	1.217	.808	1.299
70	.879	1.163	.858	1.198	.820	1.272
80	.886	1.151	.866	1.183	.829	1.250
90	.892	1.141	.873	1.171	.838	1.233
100	.897	1.133	.879	1.161	.845	1.219
>100	$\dfrac{1}{1 \pm 1.645/\sqrt{2f}}$		$\dfrac{1}{1 \pm 1.960/\sqrt{2f}}$		$\dfrac{1}{1 \pm 2.576/\sqrt{2f}}$	

* Use explained in Sec. 3.2.3. To obtain the limits, multiply sample standard deviation s by b_1 and by b_2. This table was prepared by Eleanor G. Crow, NOTS.

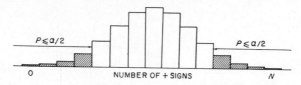

$P \leqslant \alpha/2$ $P \leqslant \alpha/2$

0 NUMBER OF + SIGNS N

TABLE 9. CRITICAL VALUES FOR THE SIGN TEST *

N \ α	.01	.05	.10	.25	N \ α	.01	.05	.10	.25
1	....	....	....	....	46	13	15	16	18
2	....	....	....	....	47	14	16	17	19
3	....	....	....	0	48	14	16	17	19
4	....	....	....	0	49	15	17	18	19
5	....	....	0	0	50	15	17	18	20
6	....	0	0	1	51	15	18	19	20
7	....	0	0	1	52	16	18	19	21
8	0	0	1	1	53	16	18	20	21
9	0	1	1	2	54	17	19	20	22
10	0	1	1	2	55	17	19	20	22
11	0	1	2	3	56	17	20	21	23
12	1	2	2	3	57	18	20	21	23
13	1	2	3	3	58	18	21	22	24
14	1	2	3	4	59	19	21	22	24
15	2	3	3	4	60	19	21	23	25
16	2	3	4	5	61	20	22	23	25
17	2	4	4	5	62	20	22	24	25
18	3	4	5	6	63	20	23	24	26
19	3	4	5	6	64	21	23	24	26
20	3	5	5	6	65	21	24	25	27
21	4	5	6	7	66	22	24	25	27
22	4	5	6	7	67	22	25	26	28
23	4	6	7	8	68	22	25	26	28
24	5	6	7	8	69	23	25	27	29
25	5	7	7	9	70	23	26	27	29
26	6	7	8	9	71	24	26	28	30
27	6	7	8	10	72	24	27	28	30
28	6	8	9	10	73	25	27	28	31
29	7	8	9	10	74	25	28	29	31
30	7	9	10	11	75	25	28	29	32
31	7	9	10	11	76	26	28	30	32
32	8	9	10	12	77	26	29	30	32
33	8	10	11	12	78	27	29	31	33
34	9	10	11	13	79	27	30	31	33
35	9	11	12	13	80	28	30	32	34
36	9	11	12	14	81	28	31	32	34
37	10	12	13	14	82	28	31	33	35
38	10	12	13	14	83	29	32	33	35
39	11	12	13	15	84	29	32	33	36
40	11	13	14	15	85	30	32	34	36
41	11	13	14	16	86	30	33	34	37
42	12	14	15	16	87	31	33	35	37
43	12	14	15	17	88	31	34	35	38
44	13	15	16	17	89	31	34	36	38
45	13	15	16	18	90	32	35	36	39

° Use explained in Sec. 2.5.2a. In the sign test the test statistic is the number of $+$ signs or the number of $-$ signs, whichever is smaller. A statistic less than or equal to the critical value tabled under α will occur with probability not more than α. For values of N larger than 90, approximate critical values may be found by taking the nearest integer less than $(N-1)/2 - k\sqrt{N+1}$, where k is 1.2879, 0.9800, 0.8224, and 0.5752 for the 1, 5, 10, and 25% values, respectively. Reproduced, with the permission of the authors and the publisher, from W. J. Dixon and F. J. Massey, Jr., *Introduction to Statistical Analysis*, New York, McGraw-Hill, 1951, p. 324.

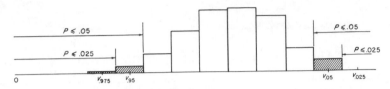

TABLE 10. CRITICAL VALUES FOR RUNS *

$v_{.975}$

n_1 / n_2	2	3	4	5	6	7	8	9	10	11	12	13	14	15	16	17	18	19	20
2	....	....	....	....	....	....	....	....	....	....	....	....	....	....	....	....	....	....	....
3	....	....	....	....	....	....	....	....	....	....	....	....	....	....	....	....	....	....	....
4	....	....	....	....	....	....	....	....	....	....	....	....	....	....	....	....	....	....	....
5	....	....	2	2	....	....	....	....	....	....	....	....	....	....	....	....	....	....	....
6	....	2	2	3	3	....	....	....	....	....	....	....	....	....	....	....	....	....	....
7	....	2	2	3	3	3	....	....	....	....	....	....	....	....	....	....	....	....	....
8	....	2	3	3	3	4	4	....	....	....	....	....	....	....	....	....	....	....	....
9	....	2	3	3	4	4	5	5	....	....	....	....	....	....	....	....	....	....	....
10	....	2	3	3	4	5	5	5	6	....	....	....	....	....	....	....	....	....	....
11	....	2	3	4	4	5	5	6	6	7	....	....	....	....	....	....	....	....	....
12	2	2	3	4	4	5	6	6	7	7	7	....	....	....	....	....	....	....	....
13	2	2	3	4	5	5	6	6	7	7	8	8	....	....	....	....	....	....	....
14	2	2	3	4	5	5	6	7	7	8	8	9	9	....	....	....	....	....	....
15	2	3	3	4	5	6	6	7	7	8	8	9	9	10	....	....	....	....	....
16	2	3	4	4	5	6	6	7	8	8	9	9	10	10	11	....	....	....	....
17	2	3	4	4	5	6	7	7	8	9	9	10	10	11	11	11	....	....	....
18	2	3	4	5	5	6	7	8	8	9	9	10	10	11	11	12	12	....	....
19	2	3	4	5	6	6	7	8	8	9	10	10	11	11	12	12	13	13	....
20	2	3	4	5	6	6	7	8	9	9	10	10	11	12	12	13	13	13	14

$n_1 = n_2$	$v_{.975}$	$n_1 = n_2$	$v_{.975}$	$n_1 = n_2$	$v_{.975}$
21	15	32	24	55	45
22	16	34	26	60	49
23	16	36	28	65	54
24	17	38	30	70	58
25	18	40	31	75	63
26	19	42	33	80	68
27	20	44	35	85	72
28	21	46	37	90	77
29	22	48	38	95	82
30	22	50	40	100	86

* Use explained in Sec. 4.1 and 4.7.2. The values listed are such that a number less than or equal to $v_{.975}$ will occur with probability not more than 2.5%; a number greater than or equal to $v_{.025}$ will will occur with probability not more than 2.5%. A number less than or equal to $v_{.95}$ will occur with probability not more than 5%; a number greater than or equal to $v_{.05}$ will occur with probability not more than 5%. Adapted, with the permission of the authors and the editor, from C. Eisenhart and F. Swed, "Tables for Testing Randomness of Grouping in a Sequence of Alternatives," ANN MATH STAT, Vol. 14 (1943), pp. 83–86.

TABLE 10. (Contd.)

$v_{.025}$

n_2 \ n_1	2	3	4	5	6	7	8	9	10	11	12	13	14	15	16	17	18	19	20
2	....	....	....	....	....	....	....	....	....	....	....	....	....	....	....	....	....	....	....
3	....	....	....	....	....	....	....	....	....	....	....	....	....	....	....	....	....	....	....
4	....	....	....	....	....	....	....	....	....	....	....	....	....	....	....	....	....	....	....
5	....	....	9	10	....	....	....	....	....	....	....	....	....	....	....	....	....	....	....
6	....	....	9	10	11	....	....	....	....	....	....	....	....	....	....	....	....	....	....
7	....	....	....	11	12	13	....	....	....	....	....	....	....	....	....	....	....	....	....
8	....	....	....	11	12	13	14	....	....	....	....	....	....	....	....	....	....	....	....
9	....	....	....	....	13	14	14	15	....	....	....	....	....	....	....	....	....	....	....
10	....	....	....	....	13	14	15	16	16	....	....	....	....	....	....	....	....	....	....
11	....	....	....	....	13	14	15	16	17	17	....	....	....	....	....	....	....	....	....
12	....	....	....	....	13	14	16	16	17	18	19	....	....	....	....	....	....	....	....
13	....	....	....	....	....	15	16	17	18	19	19	20	....	....	....	....	....	....	....
14	....	....	....	....	....	15	16	17	18	19	20	20	21	....	....	....	....	....	....
15	....	....	....	....	....	15	16	18	18	19	20	21	22	22	....	....	....	....	....
16	....	....	....	....	....	....	17	18	19	20	21	21	22	23	23	....	....	....	....
17	....	....	....	....	....	....	17	18	19	20	21	22	23	23	24	25	....	....	....
18	....	....	....	....	....	....	17	18	19	20	21	22	23	24	25	25	26	....	....
19	....	....	....	....	....	....	17	18	20	21	22	23	23	24	25	26	26	27	....
20	....	....	....	....	....	....	17	18	20	21	22	23	24	25	25	26	27	27	28

$n_1 = n_2$	$v_{.025}$	$n_1 = n_2$	$v_{.025}$	$n_1 = n_2$	$v_{.025}$
21	29	32	42	55	67
22	30	34	44	60	73
23	32	36	46	65	78
24	33	38	48	70	84
25	34	40	51	75	89
26	35	42	53	80	94
27	36	44	55	85	100
28	37	46	57	90	105
29	38	48	60	95	110
30	40	50	62	100	116

TABLE 10. (Contd.)

$v_{.95}$

n_2 \ n_1	2	3	4	5	6	7	8	9	10	11	12	13	14	15	16	17	18	19	20
2	....	....	....	....	....	....	....	....	....	....	....	....	....	....	....	....	....	....	....
3	....	....	....	....	....	....	....	....	....	....	....	....	....	....	....	....	....	....	....
4	....	....	2	....	....	....	....	....	....	....	....	....	....	....	....	....	....	....	....
5	....	2	2	3	....	....	....	....	....	....	....	....	....	....	....	....	....	....	....
6	....	2	3	3	3	....	....	....	....	....	....	....	....	....	....	....	....	....	....
7	....	2	3	3	4	4	....	....	....	....	....	....	....	....	....	....	....	....	....
8	2	2	3	3	4	4	5	....	....	....	....	....	....	....	....	....	....	....	....
9	2	2	3	4	4	5	5	6	....	....	....	....	....	....	....	....	....	....	....
10	2	3	3	4	5	5	6	6	6	....	....	....	....	....	....	....	....	....	....
11	2	3	3	4	5	5	6	6	7	7	....	....	....	....	....	....	....	....	....
12	2	3	4	4	5	6	6	7	7	8	8	....	....	....	....	....	....	....	....
13	2	3	4	4	5	6	6	7	7	8	8	9	....	....	....	....	....	....	....
14	2	3	4	5	5	6	7	7	8	8	9	9	10	....	....	....	....	....	....
15	2	3	4	5	6	6	7	8	8	9	9	10	10	11	....	....	....	....	....
16	2	3	4	5	6	6	7	8	8	9	10	10	11	11	11	....	....	....	....
17	2	3	4	5	6	7	8	8	9	9	10	10	11	11	12	12	....	....	....
18	2	3	4	5	6	7	8	8	9	10	10	11	11	12	12	13	13	....	....
19	2	3	4	5	6	7	8	8	9	10	10	11	12	12	13	13	14	14	....
20	2	3	4	5	6	7	8	9	9	10	11	11	12	12	13	13	14	14	15

$n_1 = n_2$	$v_{.95}$	$n_1 = n_2$	$v_{.95}$	$n_1 = n_2$	$v_{.95}$
21	16	32	25	55	46
22	17	34	27	60	51
23	17	36	29	65	56
24	18	38	31	70	60
25	19	40	33	75	65
26	20	42	35	80	70
27	21	44	36	85	74
28	22	46	38	90	79
29	23	48	40	95	84
30	24	50	42	100	88

TABLE 10. (Contd.)

$v_{.05}$

n_1 \ n_2	2	3	4	5	6	7	8	9	10	11	12	13	14	15	16	17	18	19	20
2	....	....	....	....	....	....	....	....	....	....	....	....	....	....	....	....	....	....	....
3	....	....	....	....	....	....	....	....	....	....	....	....	....	....	....	....	....	....	....
4	....	7	8	....	....	....	....	....	....	....	....	....	....	....	....	....	....	....	....
5	....	....	9	9	....	....	....	....	....	....	....	....	....	....	....	....	....	....	....
6	....	....	9	10	11	....	....	....	....	....	....	....	....	....	....	....	....	....	....
7	....	....	9	10	11	12	....	....	....	....	....	....	....	....	....	....	....	....	....
8	....	....	....	11	12	13	13	....	....	....	....	....	....	....	....	....	....	....	....
9	....	....	....	11	12	13	14	14	....	....	....	....	....	....	....	....	....	....	....
10	....	....	....	11	12	13	14	15	16	....	....	....	....	....	....	....	....	....	....
11	....	....	....	....	13	14	15	15	16	17	....	....	....	....	....	....	....	....	....
12	....	....	....	....	13	14	15	16	17	17	18	....	....	....	....	....	....	....	....
13	....	....	....	....	13	14	15	16	17	18	18	19	....	....	....	....	....	....	....
14	....	....	....	....	13	14	16	17	17	18	19	20	20	....	....	....	....	....	....
15	....	....	....	....	....	15	16	17	18	19	19	20	21	21	....	....	....	....	....
16	....	....	....	....	....	15	16	17	18	19	20	21	21	22	23	....	....	....	....
17	....	....	....	....	....	15	16	17	18	19	20	21	22	22	23	24	....	....	....
18	....	....	....	....	....	15	16	18	19	20	21	21	22	23	24	24	25	....	....
19	....	....	....	....	....	15	16	18	19	20	21	22	23	23	24	25	25	26	....
20	....	....	....	....	....	....	17	18	19	20	21	22	23	24	25	25	26	27	27

$n_1 = n_2$	$v_{.05}$	$n_1 = n_2$	$v_{.05}$	$n_1 = n_2$	$v_{.05}$
21	28	32	41	55	66
22	29	34	43	60	71
23	31	36	45	65	76
24	32	38	47	70	82
25	33	40	49	75	87
26	34	42	51	80	92
27	35	44	54	85	98
28	36	46	56	90	103
29	37	48	58	95	108
30	38	50	60	100	114

TABLE 11. HALF-WIDTHS d_α FOR CONSTRUCTION OF CONFIDENCE BANDS FOR CUMULATIVE DISTRIBUTIONS *

n \ α	.20	.15	.10	.05	.01
5	.45	.47	.51	.56	.67
10	.32	.34	.37	.41	.49
20	.23	.25	.26	.29	.36
25	.21	.22	.24	.27	.32
30	.19	.20	.22	.24	.29
35	.18	.19	.20	.23	.27
40	.17	.18	.19	.21	.25
45	.16	.17	.18	.20	.24
50	.15	.16	.17	.19	.23
> 50	$\dfrac{1.07}{\sqrt{n}}$	$\dfrac{1.14}{\sqrt{n}}$	$\dfrac{1.22}{\sqrt{n}}$	$\dfrac{1.36}{\sqrt{n}}$	$\dfrac{1.63}{\sqrt{n}}$

° Use explained in Sec. 4.4. Reproduced, with the permission of the authors and the publisher, from W. J. Dixon and F. J. Massey, Jr., *Introduction to Statistical Analysis*, New York, McGraw-Hill, 1951, p. 348.

TABLE 12. FACTOR k FOR ESTIMATING NORMAL STANDARD DEVIATION FROM THE RANGE w AS kw *

n	k	n	k
2	.886	8	.351
3	.591	9	.337
4	.486	10	.325
5	.430	50	.222
6	.395	100	.199
7	.370	1000	.154

° Use explained in Sec. 1.1.4b. The factors for $n = 2, 3, \ldots, 10$ are reproduced, with the permission of the authors and the publisher, from W. J. Dixon and F. J. Massey, Jr., *Introduction to Statistical Analysis*, New York, McGraw-Hill, 1951, p. 315. The factors for $n = 50, 100$, and 1000 are reciprocals of entries in a table of mean ranges complete to $n = 1000$ in L. H. C. Tippett, "On the Extreme Individuals and the Range of Samples Taken From a Normal Population," BIOMETRIKA, Vol. 17 (1925), pp. 364–87.

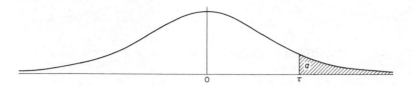

TABLE 13. CRITICAL VALUES FOR TESTS USING THE RANGE w *

(a) CRITICAL VALUES FOR $\tau_1 = \dfrac{\bar{x} - a}{w}$

n \ α	.05	.025	.01	.005
2	3.16	6.35	15.91	31.83
3	0.885	1.30	2.11	3.01
4	.529	0.717	1.02	1.32
5	.388	.507	0.685	0.843
6	.312	.399	.523	.628
7	.263	.333	.429	.507
8	.230	.288	.366	.429
9	.205	.255	.322	.374
10	.186	.230	.288	.333

(b) CRITICAL VALUES FOR $\tau_d = \dfrac{\bar{x}_1 - \bar{x}_2}{w_1 + w_2}$

$n_1 = n_2$ \ α	.05	.025	.01	.005
2	1.16	1.71	2.78	3.96
3	0.487	0.636	0.857	1.05
4	.322	.406	.524	0.618
5	.246	.306	.386	.448
6	.203	.250	.310	.357
7	.173	.213	.263	.300
8	.153	.186	.229	.260
9	.137	.167	.204	.232
10	.125	.152	.185	.210

° Use explained in Sec. 2.2.1 and 2.5.3a respectively. Adapted, with the permission of the author and the publisher, from E. Lord, "The Use of the Range in Place of the Standard Deviation in the t Test," BIOMETRIKA, Vol. 34 (1947), p. 41.

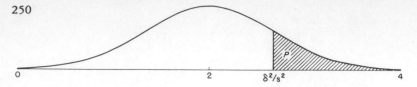

TABLE 14. Critical Values for δ^2/s^2 *

For $P = .50$, $\delta^2/s^2 = 2$ for all k.

k \ P	.999	.99	.95	.05	.01	.001
4	.5898	.6256	.7805	3.2195	3.3744	3.4102
5	.4161	.5379	.8204	3.1796	3.4621	3.5839
6	.3634	.5615	.8902	3.1098	3.4385	3.6366
7	.3695	.6140	.9359	3.0641	3.3860	3.6305
8	.4036	.6628	.9825	3.0175	3.3372	3.5964
9	.4420	.7088	1.0244	2.9756	3.2911	3.5580
10	.4816	.7518	1.0623	2.9378	3.2482	3.5184
11	.5197	.7915	1.0965	2.9035	3.2085	3.4803
12	.5557	.8280	1.1276	2.8724	3.1720	3.4443
13	.5898	.8618	1.1558	2.8442	3.1381	3.4102
14	.6223	.8931	1.1816	2.8184	3.1068	3.3777
15	.6532	.9221	1.2053	2.7947	3.0778	3.3467
16	.6826	.9491	1.2272	2.7728	3.0509	3.3174
17	.7104	.9743	1.2473	2.7527	3.0257	3.2896
18	.7368	.9979	1.2660	2.7340	3.0021	3.2632
19	.7617	1.0199	1.2834	2.7166	2.9800	3.2383
20	.7852	1.0406	1.2996	2.7004	2.9593	3.2148
21	.8073	1.0601	1.3148	2.6852	2.9399	3.1927
22	.8283	1.0785	1.3290	2.6710	2.9216	3.1718
23	.8481	1.0958	1.3425	2.6576	2.9042	3.1520
24	.8668	1.1122	1.3552	2.6449	2.8877	3.1333
25	.8846	1.1278	1.3671	2.6329	2.8722	3.1154
26	.9017	1.1426	1.3785	2.6215	2.8575	3.0983
27	.9182	1.1567	1.3892	2.6108	2.8434	3.0818
28	.9341	1.1702	1.3994	2.6006	2.8300	3.0659
29	.9496	1.1830	1.4091	2.5909	2.8171	3.0505
30	.9645	1.1951	1.4183	2.5817	2.8049	3.0355
31	.9789	1.2067	1.4270	2.5729	2.7933	3.0212
32	.9925	1.2177	1.4354	2.5646	2.7823	3.0076
33	1.0055	1.2283	1.4434	2.5566	2.7717	2.9946
34	1.0180	1.2386	1.4511	2.5490	2.7614	2.9821
35	1.0300	1.2485	1.4585	2.5416	2.7515	2.9701
36	1.0416	1.2581	1.4656	2.5344	2.7419	2.9584
37	1.0529	1.2673	1.4726	2.5274	2.7325	2.9469
38	1.0639	1.2763	1.4793	2.5208	2.7237	2.9360
39	1.0746	1.2850	1.4858	2.5142	2.7151	2.9254
40	1.0850	1.2934	1.4921	2.5079	2.7066	2.9151
41	1.0950	1.3017	1.4982	2.5018	2.6983	2.9050
42	1.1048	1.3096	1.5041	2.4958	2.6904	2.8952
43	1.1142	1.3172	1.5098	2.4901	2.6827	2.8858
44	1.1233	1.3246	1.5154	2.4846	2.6754	2.8767
45	1.1320	1.3317	1.5206	2.4794	2.6683	2.8680
46	1.1404	1.3387	1.5257	2.4743	2.6614	2.8596
47	1.1484	1.3453	1.5305	2.4695	2.6548	2.8516
48	1.1561	1.3515	1.5351	2.4649	2.6486	2.8439
49	1.1635	1.3573	1.5395	2.4604	2.6426	2.8365
50	1.1705	1.3629	1.5437	2.4563	2.6370	2.8295
51	1.1774	1.3683	1.5477	2.4522	2.6316	2.8225
52	1.1843	1.3738	1.5518	2.4483	2.6262	2.8157
53	1.1910	1.3792	1.5557	2.4444	2.6208	2.8090
54	1.1976	1.3846	1.5596	2.4405	2.6154	2.8024
55	1.2041	1.3899	1.5634	2.4368	2.6102	2.7959
56	1.2104	1.3949	1.5670	2.4330	2.6050	2.7896
57	1.2166	1.3999	1.5707	2.4294	2.6001	2.7834
58	1.2227	1.4048	1.5743	2.4258	2.5952	2.7773
59	1.2288	1.4096	1.5779	2.4222	2.5903	2.7712
60	1.2349	1.4144	1.5814	2.4186	2.5856	2.7651

* Use explained in Sec. 2.8.2. From a paper by A. H. J. Baines, "Methods of Detecting Non-Randomness in a Given Series of Observations," Technical Report — Series "R," No. Q.C./R/12, British Ministry of Supply, as reproduced in Frankford Arsenal, *Statistical Manual: Methods of Making Experimental Inferences*, 2nd rev. ed., by C. W. Churchman, Philadelphia, Frankford Arsenal, June 1951, Table VIII.

TABLE 15. ORTHOGONAL POLYNOMIALS *

n	13					14					15				
	ξ'_1	ξ'_2	ξ'_3	ξ'_4	ξ'_5	ξ'_1	ξ'_2	ξ'_3	ξ'_4	ξ'_5	ξ'_1	ξ'_2	ξ'_3	ξ'_4	ξ'_5
	0	−14	0	+84	0	+1	−8	−24	+108	+60	0	−56	0	+756	0
	+1	−13	−4	+64	+20	+3	−7	−67	+63	+145	+1	−53	−27	+621	+675
	+2	−10	−7	+11	+26	+5	−5	−95	−13	+139	+2	−44	−49	+251	+1000
	+3	−5	−8	−54	+11	+7	−2	−98	−92	+28	+3	−29	−61	−249	+751
	+4	+2	−6	−96	−18	+9	+2	−66	−132	−132	+4	−8	−58	−704	−44
	+5	+11	0	−66	−33	+11	+7	+11	−77	−187	+5	+19	−35	−869	−979
	+6	+22	+11	+99	+22	+13	+13	+143	+143	+143	+6	+52	+13	−429	−1144
											+7	+91	+91	+1001	+1001
$\Sigma\xi'^2_j$	182	2,002	572	68,068	6,188	910	728	97,240	136,136	235,144	280	37,128	39,780	6,466,460	10,581,480
λ_j	1	1	1/6	7/12	7/120	2	1/2	5/3	7/12	7/30	1	3	5/6	35/12	21/20

n	16					17					18				
	ξ'_1	ξ'_2	ξ'_3	ξ'_4	ξ'_5	ξ'_1	ξ'_2	ξ'_3	ξ'_4	ξ'_5	ξ'_1	ξ'_2	ξ'_3	ξ'_4	ξ'_5
	+1	−21	−63	+189	+45	0	−24	0	+36	0	+1	−40	−8	+44	+220
	+3	−19	−179	+129	+115	+1	−23	−7	+31	+55	+3	−37	−23	+33	+583
	+5	−15	−265	+23	+131	+2	−20	−13	+17	+88	+5	−31	−35	+13	+733
	+7	−9	−301	−101	+77	+3	−15	−17	−3	+83	+7	−22	−42	−12	+588
	+9	−1	−267	−201	−33	+4	−8	−18	−24	+36	+9	−10	−42	−36	+156
	+11	+9	−143	−221	−143	+5	+1	−15	−39	−39	+11	+5	−33	−51	−429
	+13	+21	+91	−91	−143	+6	+12	−7	−39	−104	+13	+23	−13	−47	−871
	+15	+35	+455	+273	+143	+7	+25	+7	−13	−91	+15	+44	+20	−12	−676
						+8	+40	+28	+52	+104	+17	+68	+68	+68	+884
$\Sigma\xi'^2_j$	1,360	5,712	1,007,760	470,288	201,552	408	7,752	3,876	16,796	100,776	1,938	23,256	23,256	28,424	6,953,544
λ_j	2	1	10/3	7/12	7/10	1	1	1/6	1/12	1/20	2	3/2	1/3	1/12	3/10

* Use explained in Sec. 6.3.2. Reproduced, with the permission of the authors and the publishers, from R. A. Fisher and F. Yates, *Statistical Tables*, 4th rev. ed., Edinburgh, Oliver & Boyd, Ltd., 1953. This table is an excerpt from Fisher and Yates' Table XXIII which covers values for n from 3 to 75. Further values are covered in R. L. Anderson and E. E. Houseman, "Tables of Orthogonal Polynomial Values Extended to N = 104," Iowa State College Agr. Expt. Sta. Research Bull. 297 (1942).

252

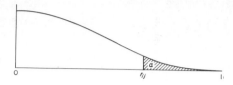

TABLE 16. CRITICAL VALUES FOR TEST RATIOS FOR GROSS ERRORS *

Ratio	Number n of observations	Critical values		
		$\alpha = .10$	$\alpha = .05$	$\alpha = .01$
$r_{10} = \dfrac{x_2 - x_1}{x_n - x_1}$	3	.941	.970	.994
	4	.765	.829	.926
	5	.642	.710	.821
	6	.560	.628	.740
	7	.507	.569	.680
$r_{11} = \dfrac{x_2 - x_1}{x_{n-1} - x_1}$	8	.544	.608	.717
	9	.503	.564	.672
	10	.470	.530	.635
	11	.445	.502	.605
	12	.423	.479	.579
$r_{22} = \dfrac{x_3 - x_1}{x_{n-2} - x_1}$	13	.563	.611	.697
	14	.539	.586	.670
	15	.518	.565	.647
	16	.500	.546	.627
	17	.483	.529	.610
	18	.469	.514	.594
	19	.457	.501	.580
	20	.446	.489	.567
	21	.435	.478	.555
	22	.426	.468	.544
	23	.418	.459	.535
	24	.410	.451	.526
	25	.402	.443	.517
	26	.396	.436	.510
	27	.389	.429	.502
	28	.383	.423	.495
	29	.378	.417	.489
	30	.373	.412	.483

* Use explained in Sec. 4.8. Values for $\alpha = .01$ and .10, $n = 3, 4, 5, 6$, and 7 are adapted from W. J. Dixon, "Ratios Involving Extreme Values," ANN MATH STAT, Vol. 22 (1951), pp. 68–78. Remaining values calculated by R. S. Gardner, NOTS. The above values apply when deviations at either end of the sample are of interest, and it is not specified before seeing the data which end it may be; Dixon's tables apply when it is known before seeing the data which end of the sample is of interest.

TABLE 17. Tolerance Factors K for Population Proportion P of Normal Distributions *

n / P	$\alpha = 0.10$ 0.90	0.95	$\alpha = 0.05$ 0.90	0.95	n / P	$\alpha = 0.10$ 0.90	0.95	$\alpha = 0.05$ 0.90	0.95
					55	1.901	2.265	1.976	2.354
2	15.978	18.800	32.019	37.674	60	1.887	2.248	1.958	2.333
3	5.847	6.919	8.380	9.916	65	1.875	2.235	1.943	2.315
4	4.166	4.943	5.369	6.370	70	1.865	2.222	1.929	2.299
5	3.494	4.152	4.275	5.079	75	1.856	2.211	1.917	2.285
6	3.131	3.723	3.712	4.414	80	1.848	2.202	1.907	2.272
7	2.902	3.452	3.369	4.007	85	1.841	2.193	1.897	2.261
8	2.743	3.264	3.136	3.732	90	1.834	2.185	1.889	2.251
9	2.626	3.125	2.967	3.532	95	1.828	2.178	1.881	2.241
10	2.535	3.018	2.839	3.379	100	1.822	2.172	1.874	2.233
11	2.463	2.933	2.737	3.259	110	1.813	2.160	1.861	2.218
12	2.404	2.863	2.655	3.162	120	1.804	2.150	1.850	2.205
13	2.355	2.805	2.587	3.081	130	1.797	2.141	1.841	2.194
14	2.314	2.756	2.529	3.012	140	1.791	2.134	1.833	2.184
15	2.278	2.713	2.480	2.954	150	1.785	2.127	1.825	2.175
16	2.246	2.676	2.437	2.903	160	1.780	2.121	1.819	2.167
17	2.219	2.643	2.400	2.858	170	1.775	2.116	1.813	2.160
18	2.194	2.614	2.366	2.819	180	1.771	2.111	1.808	2.154
19	2.172	2.588	2.337	2.784	190	1.767	2.106	1.803	2.148
20	2.152	2.564	2.310	2.752	200	1.764	2.102	1.798	2.143
21	2.135	2.543	2.286	2.723	225	1.756	2.093	1.788	2.131
22	2.118	2.524	2.264	2.697	250	1.750	2.085	1.780	2.121
23	2.103	2.506	2.244	2.673	275	1.745	2.079	1.773	2.113
24	2.089	2.489	2.225	2.651	300	1.740	2.073	1.767	2.106
25	2.077	2.474	2.208	2.631					
26	2.065	2.460	2.193	2.612	350	1.732	2.064	1.757	2.094
27	2.054	2.447	2.178	2.595	400	1.726	2.057	1.749	2.084
28	2.044	2.435	2.164	2.579	450	1.721	2.051	1.743	2.077
29	2.034	2.424	2.152	2.564	500	1.717	2.046	1.737	2.070
30	2.025	2.413	2.140	2.549					
31	2.017	2.403	2.129	2.536	550	1.713	2.041	1.733	2.065
32	2.009	2.393	2.118	2.524	600	1.710	2.038	1.729	2.060
33	2.001	2.385	2.108	2.512	650	1.707	2.034	1.725	2.056
34	1.994	2.376	2.099	2.501	700	1.705	2.032	1.722	2.052
35	1.988	2.368	2.090	2.490	750	1.703	2.029	1.719	2.049
36	1.981	2.361	2.081	2.479	800	1.701	2.027	1.717	2.046
37	1.975	2.353	2.073	2.470	850	1.699	2.025	1.714	2.043
38	1.969	2.346	2.068	2.464	900	1.697	2.023	1.712	2.040
39	1.964	2.340	2.060	2.455	950	1.696	2.021	1.710	2.038
40	1.959	2.334	2.052	2.445	1000	1.695	2.019	1.709	2.036
45	1.935	2.306	2.021	2.408	∞	1.645	1.960	1.645	1.960
50	1.916	2.284	1.996	2.379					

* Use explained in Sec. 4.9. Abridged, with the permission of the author and the publisher, from A. H. Bowker, "Tolerance Limits for Normal Distributions," in *Selected Techniques of Statistical Analysis,* ed. by Eisenhart, Hastay, and Wallis, New York, McGraw-Hill, 1947, pp. 97–110. The original reference contains five P values for each of four α values and includes more n values.

TABLE 18. FACTORS FOR COMPUTING CONTROL-CHART LINES *

Sample size n	Chart for averages				Chart for standard deviations						Chart for ranges					
	Factors for control limits				Factors for central line		Factors for control limits				Factors for central line		Factors for control limits			
	A	A_0	A_1	A_2	c_2	$1/c_2$	B_1	B_2	B_3	B_4	d_2	$1/d_2$	D_1	D_2	D_3	D_4
2	2.121	3.760	3.760	1.880	0.5642	1.7725	0	1.843	0	3.267	1.128	0.8865	0	3.686	0	3.267
3	1.732	3.070	2.394	1.023	0.7236	1.3820	0	1.858	0	2.568	1.693	0.5907	0	4.358	0	2.575
4	1.500	2.914	1.880	0.729	0.7979	1.2533	0	1.808	0	2.266	2.059	0.4857	0	4.698	0	2.282
5	1.342	2.884	1.596	0.577	0.8407	1.1894	0	1.756	0	2.089	2.326	0.4299	0	4.918	0	2.115
6	1.225	2.899	1.410	0.483	0.8686	1.1512	0.026	1.711	0.030	1.970	2.534	0.3946	0	5.078	0	2.004
7	1.134	2.935	1.277	0.419	0.8882	1.1259	0.105	1.672	0.118	1.882	2.704	0.3698	0.205	5.203	0.076	1.924
8	1.061	2.980	1.175	0.373	0.9027	1.1078	0.167	1.638	0.185	1.815	2.847	0.3512	0.387	5.307	0.136	1.864
9	1.000	3.030	1.094	0.337	0.9139	1.0942	0.219	1.609	0.239	1.761	2.970	0.3367	0.546	5.394	0.184	1.816
10	0.949	3.085	1.028	0.308	0.9227	1.0837	0.262	1.584	0.284	1.716	3.078	0.3249	0.687	5.469	0.223	1.777
11	0.905	3.136	0.973	0.285	0.9300	1.0753	0.299	1.561	0.321	1.679	3.173	0.3152	0.812	5.534	0.256	1.744
12	0.866	3.189	0.925	0.266	0.9359	1.0684	0.331	1.541	0.354	1.646	3.258	0.3069	0.924	5.592	0.284	1.716
13	0.832	3.242	0.884	0.249	0.9410	1.0627	0.359	1.523	0.382	1.618	3.336	0.2998	1.026	5.646	0.308	1.692
14	0.802	3.295	0.848	0.235	0.9453	1.0579	0.384	1.507	0.406	1.594	3.407	0.2935	1.121	5.693	0.329	1.671
15	0.775	3.347	0.816	0.223	0.9490	1.0537	0.406	1.492	0.428	1.572	3.472	0.2880	1.207	5.737	0.348	1.652
16	0.750	3.398	0.788	0.212	0.9523	1.0501	0.427	1.478	0.448	1.552	3.532	0.2831	1.285	5.779	0.364	1.636
17	0.728	3.448	0.762	0.203	0.9551	1.0470	0.445	1.465	0.466	1.534	3.588	0.2787	1.359	5.817	0.379	1.621
18	0.707	3.497	0.738	0.194	0.9576	1.0442	0.461	1.454	0.482	1.518	3.640	0.2747	1.426	5.854	0.392	1.608
19	0.688	3.545	0.717	0.187	0.9599	1.0418	0.477	1.443	0.497	1.503	3.689	0.2711	1.490	5.888	0.404	1.596
20	0.671	3.592	0.697	0.180	0.9619	1.0396	0.491	1.433	0.510	1.490	3.735	0.2677	1.548	5.922	0.414	1.586
21	0.655	3.639	0.679	0.173	0.9638	1.0376	0.504	1.424	0.523	1.477	3.778	0.2647	1.606	5.950	0.425	1.575
22	0.640	3.684	0.662	0.167	0.9655	1.0358	0.516	1.415	0.534	1.466	3.819	0.2618	1.659	5.979	0.434	1.566
23	0.626	3.729	0.647	0.162	0.9670	1.0342	0.527	1.407	0.545	1.455	3.858	0.2592	1.710	6.006	0.443	1.557
24	0.612	3.773	0.632	0.157	0.9684	1.0327	0.538	1.399	0.555	1.445	3.895	0.2567	1.759	6.031	0.452	1.548
25	0.600	3.816	0.619	0.153	0.9696	1.0313	0.548	1.392	0.565	1.435	3.931	0.2544	1.804	6.058	0.459	1.541
>25	$\dfrac{3}{\sqrt{n}}$		$\dfrac{3}{\sqrt{n}}$		1.0000	1.0000	**	***	**	***						

* Use explained in Chapter 7. The relation $A_0 = 3\sqrt{n}/d_2$ holds. Adapted, with the permission of the American Society for Testing Materials, from ASTM Manual on Quality Control of Materials, Philadelphia, January 1951, p. 115.

** $1 - \dfrac{3}{\sqrt{2n}}$

*** $1 + \dfrac{3}{\sqrt{2n}}$

TABLE 19. CONSTANTS FOR A SAMPLING PLAN WITH OC CURVE
THROUGH $(p_1, 0.95)$ AND $(p_2, 0.05)$ *

(a) SINGLE SAMPLING PLAN

Operating ratio $R_0 = p_2/p_1$	Ac	np_1
58.	0	0.051
13.	1	0.355
7.7	2	0.818
5.7	3	1.366
4.6	4	1.970
4.0	5	2.61
3.6	6	3.29
3.3	7	3.98
3.1	8	4.70
2.9	9	5.43
2.7	10	6.17
2.5	12	7.69
2.37	14	9.25
2.03	21	14.89
1.81	30	24.44
1.61	47	37.20
1.51	63	51.43
1.335	129	111.83
1.251	215	192.41

(b) DOUBLE SAMPLING PLAN

Operating ratio $R_0 = p_2/p_1$	Ac_1	Re_1	Ac_2	n_1p_1
15.1	0	2	1	0.207
8.3	0	3	2	0.427
5.1	1	4	4	1.00
4.1	2	5	6	1.63
3.5	2	6	7	1.99
3.0	3	8	9	2.77
2.6	5	12	13	4.34
2.3	6	14	16	5.51
2.02	9	18	23	8.38
1.82	13	24	32	12.19
1.61	21	35	50	20.04
1.505	30	46	69	28.53
1.336	63	84	138	60.31

* Use explained in Sec. 8.3. Reproduced with the permission of the author and the publishers from Paul Peach, *An Introduction to Industrial Statistics and Quality Control*, 2nd ed., Raleigh, N. C., Edwards & Broughton Co., 1947, pp. 228–29.

TABLE 20. TABLE OF $a = \log_{10}\dfrac{1-\beta}{\alpha}$ AND $b = \log_{10}\dfrac{1-\alpha}{\beta}$

The upper number in each cell represents a, the lower number, b.

β \\ α	0.001	0.010	0.050	0.100	0.200
0.001	3.000 3.000	2.000 2.996	1.301 2.978	1.000 2.954	0.699 2.903
0.010	2.996 2.000	1.996 1.996	1.297 1.978	0.996 1.954	0.695 1.903
0.050	2.978 1.301	1.978 1.297	1.279 1.279	0.978 1.255	0.677 1.204
0.100	2.954 1.000	1.954 0.996	1.255 0.978	0.954 0.954	0.653 0.903
0.200	2.903 0.699	1.903 0.695	1.204 0.677	0.903 0.653	0.602 0.602

TABLE 21. CONFIDENCE LIMITS FOR A PROPORTION

(a) TWO-SIDED LIMITS*

Upper limits are in **boldface**. The observed proportion in a random sample is r/n.

r	90%		95%		99%		r	90%		95%		99%	
			n = 1							**n = 2**			
0	0	**.900**	0	**.950**	0	**.990**	0	0	**.684**	0	**.776**	0	**.900**
1	.100	**1**	.050	**1**	.010	**1**	1	.051	**.949**	.025⁺	**.975⁻**	.005⁺	**.995⁻**
							2	.316	**1**	.224	**1**	.100	**1**
			n = 3							**n = 4**			
0	0	**.536**	0	**.632**	0	**.785⁻**	0	0	**.500**	0	**.527**	0	**.684**
1	.035⁻	**.804**	.017	**.865⁻**	.003	**.941**	1	.026	**.680**	.013	**.751**	.003	**.859**
2	.196	**.965⁺**	.135⁺	**.983**	.059	**.997**	2	.143	**.857**	.098	**.902**	.042	**.958**
3	.464	**1**	.368	**1**	.215⁺	**1**	3	.320	**.974**	.249	**.987**	.141	**.997**
							4	.500	**1**	.473	**1**	.316	**1**
			n = 5							**n = 6**			
0	0	**.379**	0	**.500**	0	**.602**	0	0	**.345⁻**	0	**.402**	0	**.536**
1	.021	**.621**	.010	**.657**	.002	**.778**	1	.017	**.542**	.009	**.598**	.002	**.706**
2	.112	**.753**	.076	**.811**	.033	**.894**	2	.093	**.667**	.063	**.729**	.027	**.827**
3	.247	**.888**	.189	**.924**	.106	**.967**	3	.201	**.799**	.153	**.847**	.085⁺	**.915⁺**
4	.379	**.979**	.343	**.990**	.222	**.998**	4	.333	**.907**	.271	**.937**	.173	**.973**
5	.621	**1**	.500	**1**	.398	**1**	5	.458	**.983**	.402	**.991**	.294	**.998**
							6	.655⁺	**1**	.598	**1**	.464	**1**
			n = 7							**n = 8**			
0	0	**.316**	0	**.377**	0	**.500**	0	0	**.255⁻**	0	**.315⁺**	0	**.451**
1	.015⁻	**.500**	.007	**.554**	.001	**.643**	1	.013	**.418**	.006	**.500**	.001	**.590**
2	.079	**.684**	.053	**.659**	.023	**.764**	2	.069	**.582**	.046	**.685⁻**	.020	**.707**
3	.170	**.721**	.129	**.775⁻**	.071	**.858**	3	.147	**.745⁺**	.111	**.711**	.061	**.802**
4	.279	**.830**	.225⁺	**.871**	.142	**.929**	4	.240	**.760**	.193	**.807**	.121	**.879**
5	.316	**.921**	.341	**.947**	.236	**.977**	5	.255⁻	**.853**	.289	**.889**	.198	**.939**
6	.500	**.985⁺**	.446	**.993**	.357	**.999**	6	.418	**.931**	.315⁺	**.954**	.293	**.980**
7	.684	**1**	.623	**1**	.500	**1**	7	.582	**.987**	.500	**.994**	.410	**.999**
							8	.745⁺	**1**	.685⁻	**1**	.549	**1**
			n = 9							**n = 10**			
0	0	**.232**	0	**.289**	0	**.402**	0	0	**.222**	0	**.267**	0	**.376**
1	.012	**.391**	.006	**.443**	.001	**.598**	1	.010	**.352**	.005⁺	**.397**	.001	**.512**
2	.061	**.515⁺**	.041	**.558**	.017	**.656**	2	.055⁻	**.500**	.037	**.603**	.016	**.624**
3	.129	**.610**	.098	**.711**	.053	**.750**	3	.116	**.648**	.087	**.619**	.048	**.703**
4	.210	**.768**	.169	**.749**	.105⁺	**.829**	4	.188	**.659**	.150	**.733**	.093	**.782**
5	.232	**.790**	.251	**.831**	.171	**...895⁻**	5	.222	**.778**	.222	**.778**	.150	**.850**
6	.390	**.871**	.289	**.902**	.250	**.947**	6	.341	**.812**	.267	**.850**	.218	**.907**
7	.485⁻	**.939**	.442	**.959**	.344	**.983**	7	.352	**.884**	.381	**.913**	.297	**.952**
8	.609	**.988**	.557	**.994**	.402	**.999**	8	.500	**.945⁺**	.397	**.963**	.376	**.984**
9	.768	**1**	.711	**1**	.598	**1**	9	.648	**.990**	.603	**.995⁻**	.488	**.999**
							10	.778	**1**	.733	**1**	.624	**1**
			n = 11							**n = 12**			
0	0	**.197**	0	**.250**	0	**.359**	0	0	**.184**	0	**.236**	0	**.321**
1	.010	**.315⁺**	.005⁻	**.369**	.001	**.500**	1	.009	**.294**	.004	**.346**	.001	**.445⁺**
2	.049	**.423**	.033	**.500**	.014	**.593**	2	.045⁺	**.398**	.030	**.450**	.013	**.555⁻**
3	.105⁻	**.577**	.079	**.631**	.043	**.660**	3	.096	**.500**	.072	**.550**	.039	**.679**
4	.169	**.685⁻**	.135⁺	**.667**	.084	**.738**	4	.154	**.602**	.123	**.654**	.076	**.698**
5	.197	**.698**	.200	**.750**	.134	**.806**	5	.184	**.706**	.181	**.706**	.121	**.765⁺**

° Use explained in Sec. 2.3.3. See Appendix Charts II, III, and IV for sample size $n > 30$. This table was calculated by Edwin L. Crow, Eleanor G. Crow, and Robert S. Gardner, NOTS, according to a modification of a proposal of Theodore E. Sterne.

TABLE 21 (Contd.)

(a) Two-Sided Limits (Contd.)

$n = 11$

r	90%		95%		99%	
6	.302	.803	.250	.800	.194	.866
7	.315+	.831	.333	.865-	.262	.916
8	.423	.895+	.369	.921	.340	.957
9	.577	.951	.500	.967	.407	.986
10	.685-	.990	.631	.995+	.500	.999
11	.803	1	.750	1	.641	1

$n = 12$

r	90%		95%		99%	
6	.271	.729	.236	.764	.175-	.825+
7	.294	.816	.294	.819	.235-	.879
8	.398	.846	.346	.877	.302	.924
9	.500	.904	.450	.928	.321	.961
10	.602	.955-	.550	.970	.445+	.987
11	.706	.991	.654	.996	.555-	.999
12	.816	1	.764	1	.679	1

$n = 13$

r	90%		95%		99%	
0	0	.173	0	.225+	0	.302
1	.008	.276	.004	.327	.001	.429
2	.042	.379	.028	.434	.012	.523
3	.088	.470	.066	.520	.036	.594
4	.142	.545-	.113	.587	.069	.698
5	.173	.621	.166	.673	.111	.727
6	.246	.724	.224	.740	.159	.787
7	.276	.754	.260	.776	.213	.841
8	.379	.827	.327	.834	.273	.889
9	.455+	.858	.413	.887	.302	.931
10	.530	.912	.480	.934	.406	.964
11	.621	.958	.566	.972	.477	.988
12	.724	.992	.673	.996	.571	.999
13	.827	1	.775-	1	.698	1

$n = 14$

r	90%		95%		99%	
0	0	.163	0	.207	0	.286
1	.007	.261	.004	.312	.001	.392
2	.039	.365+	.026	.389	.011	.500
3	.081	.422	.061	.500	.033	.608
4	.131	.578	.104	.611	.064	.636
5	.163	.594	.153	.629	.102	.714
6	.224	.645+	.206	.688	.146	.751
7	.261	.739	.207	.793	.195-	.805+
8	.355-	.776	.312	.794	.249	.854
9	.406	.837	.371	.847	.286	.898
10	.422	.869	.389	.896	.364	.936
11	.578	.919	.500	.939	.392	.967
12	.635-	.961	.611	.974	.500	.989
13	.739	.993	.688	.996	.608	.999
14	.837	1	.793	1	.714	1

$n = 15$

r	90%		95%		99%	
0	0	.154	0	.191	0	.273
1	.007	.247	.003	.302	.001	.373
2	.036	.326	.024	.369	.010	.461
3	.076	.400	.057	.448	.031	.539
4	.122	.500	.097	.552	.059	.627
5	.154	.600	.142	.631	.094	.672
6	.205+	.674	.191	.668	.135-	.727
7	.247	.675-	.192	.706	.179	.771
8	.325+	.753	.294	.808	.229	.821
9	.326	.795-	.332	.809	.273	.865+
10	.400	.846	.369	.858	.328	...906
11	.500	.878	.448	.903	.373	.941
12	.600	.924	.552	.943	.461	.969
13	.674	.964	.631	.976	.539	.990
14	.753	.993	.698	.997	.627	.999
15	.846	1	.809	1	.727	1

$n = 16$

r	90%		95%		99%	
0	0	.147	0	.179	0	.264
1	.007	.235+	.003	.273	.001	.357
2	.034	.305+	.023	.352	.010	.451
3	.071	.381	.053	.429	.029	.525-
4	.114	.450	.090	.500	.055+	.579
5	.147	.550	.132	.571	.088	.643
6	.189	.619	.178	.648	.125+	.705-
7	.235+	.695-	.179	.727	.166	.739
8	.299	.701	.272	.728	.212	.788
9	.305+	.765-	.273	.821	.261	.834
10	.381	.811	.352	.822	.295+	.875-
11	.450	.853	.429	.868	.357	.912
12	.550	.886	.500	.910	.421	.945-
13	.619	.929	.571	.947	.475+	.971
14	.695-	.966	.648	.977	.549	.990
15	.765-	.993	.727	.997	.643	.999
16	.853	1	.821	1	.736	1

$n = 17$

r	90%		95%		99%	
0	0	.140	0	.167	0	.243
1	.006	.225+	.003	.254	.001	.346
2	.032	.290	.021	.337	.009	.413
3	.067	.364	.050	.417	.027	.500
4	.107	.432	.085-	.489	.052	.587
5	.140	.500	.124	.544	.082	.620
6	.175+	.568	.166	.594	.117	.662
7	.225+	.636	.167	.663	.155+	.757
8	.277	.710	.253	.746	.197	.758
9	.290	.723	.254	.747	.242	.803
10	.364	.775-	.337	.833	.243	.845-

$n = 18$

r	90%		95%		99%	
0	0	.135-	0	.157	0	.228
1	.006	.216	.003	.242	.001	.318
2	.030	.277	.020	.325+	.008	.397
3	.063	.349	.047	.381	.025+	.466
4	.101	.419	.080	.444	.049	.534
5	.135-	.482	.116	.556	.077	.603
6	.163	.536	.156	.619	.110	.682
7	.216	.584	.157	.625+	.145+	.686
8	.257	.651	.236	.675+	.184	.772
9	.277	.723	.242	.758	.226	.774
10	.349	.743	.325-	.764	.228	.816

TABLE 21 (Contd.)

(a) Two-Sided Limits (Contd.)

r	90%		95%		99%		r	90%		95%		99%	
	n = 17							**n = 18**					
11	.432	.825⁻	.406	.834	.338	.883	11	.416	.784	.375⁻	.843	.314	.855⁻
12	.500	.860	.456	.876	.380	.918	12	.464	.837	.381	.844	.318	.890
13	.568	.893	.511	.915⁺	.413	.948	13	.518	.865⁺	.444	.884	.397	.923
14	.636	.933	.583	.950	.500	.973	14	.581	.899	.556	.920	.466	.951
15	.710	.968	.663	.979	.587	.991	15	.651	.937	.619	.953	.534	.975⁻
16	.775⁻	.994	.746	.997	.654	.999	16	.723	.970	.675⁺	.980	.603	.992
17	.860	1	.833	1	.757	1	17	.784	.994	.758	.997	.682	.999
							18	.865⁺	1	.843	1	.772	1
	n = 19							**n = 20**					
0	0	.130	0	.150	0	.218	0	0	.126	0	.143	0	.209
1	.006	.209	.003	.232	.001	.305⁺	1	.005⁺	.203	.003	.222	.001	.293
2	.028	.265⁺	.019	.316	.008	.383	2	.027	.255⁻	.018	.294	.008	.375⁻
3	.059	.337	.044	.365⁻	.024	.455⁺	3	.056	.328	.042	.351	.023	.424
4	.095⁺	.387	.075⁺	.426	.046	.515⁺	4	.090	.367	.071	.411	.044	.500
5	.130	.440	.110	.500	.073	.564	5	.126	.422	.104	.467	.069	.576
6	.151	.560	.147	.574	.103	.617	6	.141	.500	.140	.533	.098	.601
7	.209	.613	.150	.635⁺	.137	.695⁻	7	.201	.578	.143	.589	.129	.637
8	.238	.614	.222	.655⁺	.173	.707	8	.221	.633	.209	.649	.163	.707
9	.265⁺	.663	.232	.688	.212	.782	9	.255⁻	.642	.222	.706	.200	.726
10	.337	.735⁻	.312	.768	.218	.788	10	.325⁻	.675⁺	.293	.707	.209	.791
11	.386	.762	.345⁻	.778	.293	.827	11	.358	.745⁻	.294	.778	.274	.800
12	.387	.791	.365⁻	.850	.305⁺	.863	12	.367	.779	.351	.791	.293	.837
13	.440	.849	.426	.853	.383	.897	13	.422	.799	.411	.857	.363	.871
14	.560	.870	.500	.890	.436	.927	14	.500	.859	.467	.860	.399	.902
15	.613	.905⁻	.574	.925⁺	.485⁻	.954	15	.578	.874	.533	.896	.424	.931
16	.663	.941	.635⁺	.956	.545⁻	.976	16	.633	.910	.589	.929	.500	.956
17	.735⁻	.972	.684	.981	.617	.992	17	.672	.944	.649	.958	.576	.977
18	.791	.994	.768	.997	.695⁻	.999	18	.745⁺	.973	.706	.982	.625⁺	.992
19	.870	1	.850	1	.782	1	19	.797	.995⁻	.778	.997	.707	.999
							20	.874	1	.857	1	.791	1
	n = 21							**n = 22**					
0	0	.123	0	.137	0	.201	0	0	.116	0	.132	0	.194
1	.005⁺	.192	.002	.213	.000	.283	1	.005⁻	.182	.002	.205⁺	.000	.273
2	.026	.245⁻	.017	.277	.007	.347	2	.024	.236	.016	.264	.007	.334
3	.054	.307	.040	.338	.022	.409	3	.051	.289	.038	.326	.021	.396
4	.086	.353	.068	.398	.041	.466	4	.082	.340	.065⁻	.389	.039	.454
5	.121	.407	.099	.455⁺	.065⁺	.534	5	.115⁻	.393	.094	.424	.062	.505⁻
6	.130	.458	.132	.506	.092	.591	6	.116	.444	.126	.500	.088	.550
7	.191	.542	.137	.551	.122	.653	7	.181	.500	.132	.576	.116	.604
8	.192	.593	.197	.602	.155⁻	.661	8	.182	.556	.187	.582	.147	.666
9	.245⁻	.647	.213	.662	.189	.717	9	.236	.607	.205⁺	.617	.179	.682
10	.306	.693	.276	.723	.201	.743	10	.289	.660	.260	.674	.194	.727
11	.307	.694	.277	.724	.257	.799	11	.290	.710	.264	.736	.242	.758
12	.353	.755⁺	.338	.787	.283	.811	12	.340	.711	.326	.740	.273	.806
13	.407	.808	.398	.803	.339	.845⁺	13	.393	.764	.383	.795⁻	.318	.821
14	.458	.809	.449	.863	.347	.878	14	.444	.818	.418	.813	.334	.853
15	.542	.870	.494	.868	.409	.908	15	.500	.819	.424	.868	.396	.884
16	.593	.879	.545⁻	.901	.466	.935⁻	16	.556	.884	.500	.874	.450	.912
17	.647	.914	.602	.932	.534	.959	17	.607	.885⁺	.576	.906	.495⁺	.938
18	.693	.946	.662	.960	.591	.978	18	.660	.918	.611	.935⁺	.546	.961
19	.755⁺	.974	.723	.983	.653	.993	19	.711	.949	.674	.962	.604	.979
20	.808	.995⁻	.787	.998	.717	1.000	20	.764	.976	.736	.984	.666	.993
21	.877	1	.863	1	.799	1	21	.818	.995⁺	.795⁻	.998	.727	1.000
							22	.884	1	.868	1	.806	1

TABLE 21 (Contd.)

(a) Two-Sided Limits (Contd.)

n = 23 and n = 24

r	90%		95%		99%		r	90%		95%		99%	
	n = 23							n = 24					
0	0	.111	0	.127	0	.187	0	0	.105+	0	.122	0	.181
1	.005-	.174	.002	.198	.000	.265+	1	.004	.165+	.002	.191	.000	.259
2	.023	.228	.016	.255-	.007	.323	2	.022	.221	.015+	.246	.006	.313
3	.049	.274	.037	.317	.020	.386	3	.047	.264	.035-	.308	.019	.364
4	.078	.328	.062	.361	.038	.429	4	.075-	.317	.059	.347	.036	.416
5	.110	.381	.090	.409	.059	.500	5	.105+	.370	.086	.396	.057	.464
6	.111	.431	.120	.457	.084	.571	6	.105+	.423	.115-	.443	.080	.536
7	.173	.479	.127	.543	.111	.580	7	.165-	.448	.122	.500	.106	.584
8	.174	.522	.178	.591	.140	.616	8	.165+	.552	.169	.557	.133	.636
9	.228	.569	.198	.639	.171	.677	9	.221	.553	.191	.604	.163	.638
10	.273	.619	.247	.640	.187	.702	10	.259	.587	.234	.653	.181	.687
11	.274	.672	.255-	.683	.229	.735-	11	.264	.630	.246	.661	.216	.720
12	.328	.726	.317	.745+	.265+	.771	12	.317	.683	.308	.692	.257	.743
13	.381	.727	.360	.753	.298	.813	13	.370	.736	.339	.754	.280	.784
14	.431	.772	.361	.802	.323	.829	14	.413	.741	.347	.766	.313	.819
15	.478	.826	.409	.822	.384	.860	15	.447	.779	.396	.809	.362	.837
16	.521	.827	.457	.873	.420	.889	16	.448	.835-	.443	.831	.364	.867
17	.569	.889	.543	.880	.429	.916	17	.552	.835+	.500	.878	.416	.894
18	.619	.890	.591	.910	.500	.941	18	.577	.895-	.557	.885+	.464	.920
19	.672	.922	.639	.938	.571	.962	19	.630	.895+	.604	.914	.536	.943
20	.726	.951	.683	.963	.614	.980	20	.683	.925+	.653	.941	.584	.964
21	.772	.977	.745+	.984	.677	.993	21	.736	.953	.692	.965+	.636	.981
22	.826	.995+	.802	.998	.735-	1.000	22	.779	.978	.754	.985-	.687	.994
23	.889	1	.873	1	.813	1	23	.835-	.996	.809	.998	.741	1.000
							24	.895-	1	.878	1	.819	1

n = 25 and n = 26

r	90%		95%		99%		r	90%		95%		99%	
	n = 25							n = 26					
0	0	.102	0	.118	0	.175+	0	0	.098	0	.114	0	.170
1	.004	.159	.002	.185+	.000	.246	1	.004	.152	.002	.180	.000	.235-
2	.021	.214	.014	.238	.006	.305-	2	.021	.209	.014	.230	.006	.298
3	.045-	.255-	.034	.303	.018	.352	3	.043	.247	.032	.283	.017	.342
4	.072	.307	.057	.336	.034	.403	4	.069	.299	.054	.325+	.033	.393
5	.101	.362	.082	.384	.054	.451	5	.097	.343	.079	.374	.052	.442
6	.102	.390	.110	.431	.077	.500	6	.098	.377	.106	.421	.073	.487
7	.158	.432	.118	.475-	.101	.549	7	.151	.419	.114	.465-	.097	.526
8	.159	.500	.161	.525-	.127	.597	8	.152	.460	.154	.506	.122	.562
9	.214	.568	.185+	.569	.155+	.648	9	.209	.540	.180	.542	.149	.607
10	.246	.610	.222	.616	.175+	.658	10	.233	.581	.212	.579	.170	.658
11	.255-	.611	.238	.664	.205+	.695+	11	.247	.623	.230	.626	.195-	.678
12	.307	.640	.296	.683	.245+	.754	12	.299	.657	.282	.675-	.234	.702
13	.360	.693	.317	.704	.246	.755-	13	.342	.658	.283	.717	.235-	.765+
14	.389	.745+	.336	.762	.305-	.795-	14	.343	.701	.325+	.718	.298	.766
15	.390	.754	.384	.778	.342	.825-	15	.377	.753	.374	.770	.322	.805+
16	.432	.786	.431	.815-	.352	.845-	16	.419	.767	.421	.788	.342	.830
17	.500	.841	.475-	.839	.403	.873	17	.460	.791	.458	.820	.393	.851
18	.568	.842	.525+	.882	.451	.899	18	.540	.848	.494	.846	.438	.878
19	.610	.898	.569	.890	.500	.923	19	.581	.849	.535-	.886	.474	.903
20	.638	.899	.616	.918	.549	.946	20	.623	.902	.579	.894	.513	.927
21	.693	.928	.664	.943	.597	.966	21	.657	.903	.626	.921	.558	.948
22	.745+	.955+	.697	.966	.648	.982	22	.701	.931	.675-	.946	.607	.967
23	.786	.979	.762	.986	.695+	.994	23	.753	.957	.717	.968	.658	.983
24	.841	.996	.815-	.998	.754	1.000	24	.791	.979	.770	.986	.702	.994
25	.898	1	.882	1	.825-	1	25	.848	.996	.820	.998	.765+	1.000
							26	.902	1	.886	1	.830	1

TABLE 21 (Contd.)
(a) Two-Sided Limits (Contd.)

n = 27 and n = 28

r	90%		95%		99%		r	90%		95%		99%	
								n = 27				*n = 28*	
0	0	.093	0	.110	0	.166	0	0	.090	0	.106	0	.162
1	.004	.146	.002	.175⁻	.000	.225⁻	1	.004	.140	.002	.170	.000	.218
2	.020	.204	.013	.223	.006	.297	2	.019	.201	.013	.217	.005⁺	.273
3	.042	.239	.031	.270	.017	.332	3	.040	.232	.030	.259	.016	.323
4	.066	.291	.052	.316	.032	.384	4	.064	.284	.050	.307	.031	.365⁻
5	.093	.327	.076	.364	.050	.419	5	.089	.312	.073	.357	.048	.408
6	.094	.365⁺	.101	.415⁻	.070	.461	6	.090	.355⁻	.098	.384	.068	.449
7	.145⁺	.407	.110	.437	.093	.539	7	.139	.396	.106	.424	.089	.500
8	.146	.447	.148	.500	.117	.581	8	.140	.435⁺	.142	.463	.112	.551
9	.204	.500	.175⁻	.563	.143	.587	9	.197	.473	.170	.537	.137	.592
10	.221	.553	.202	.570	.166	.617	10	.208	.527	.192	.576	.162	.635⁺
11	.239	.593	.223	.598	.185⁻	.668	11	.232	.565⁻	.217	.616	.175⁺	.636
12	.291	.635⁻	.269	.636	.224	.702	12	.284	.604	.258	.619	.214	.677
13	.326	.673	.270	.684	.225⁻	.716	13	.310	.645⁺	.259	.645⁺	.218	.727
14	.327	.674	.316	.730	.284	.775⁺	14	.312	.688	.307	.693	.272	.728
15	.365⁺	.709	.364	.731	.298	.776	15	.355⁻	.690	.355⁻	.741	.273	.782
16	.407	.761	.402	.777	.332	.815⁺	16	.396	.716	.381	.742	.323	.786
17	.447	.779	.430	.798	.383	.834	17	.435⁺	.768	.384	.783	.364	.825⁻
18	.500	.796	.437	.825⁺	.413	.857	18	.473	.792	.424	.808	.365⁻	.838
19	.553	.854	.500	.852	.419	.883	19	.527	.803	.463	.830	.408	.863
20	.593	.855⁻	.563	.890	.461	.907	20	.565⁺	.860	.537	.858	.449	.888
21	.635⁻	.906	.585⁺	.899	.539	.930	21	.604	.861	.576	.894	.500	.911
22	.673	.907	.636	.924	.581	.950	22	.645⁺	.910	.616	.902	.551	.932
23	.709	.934	.684	.948	.616	.968	23	.688	.911	.643	.927	.592	.952
24	.761	.958	.730	.969	.668	.983	24	.716	.936	.693	.950	.635⁺	.969
25	.796	.980	.777	.987	.703	.994	25	.768	.960	.741	.970	.677	.984
26	.854	.996	.825⁺	.998	.775⁺	1.000	26	.799	.981	.783	.987	.727	.995⁻
27	.907	1	.890	1	.834	1	27	.860	.996	.830	.998	.782	1.000
							28	.910	1	.894	1	.838	1

n = 29 and n = 30

r	90%		95%		99%		r	90%		95%		99%	
								n = 29				*n = 30*	
0	0	.087	0	.103	0	.160	0	0	.084	0	.100	0	.152
1	.004	.135⁻	.002	.166	.000	.211	1	.004	.130	.002	.163	.000	.206
2	.018	.190	.012	.211	.005⁺	.263	2	.018	.183	.012	.205⁺	.005⁺	.256
3	.039	.225⁻	.029	.251	.015⁺	.316	3	.037	.219	.028	.244	.015⁻	.310
4	.062	.279	.049	.299	.030	.354	4	.059	.266	.047	.292	.028	.345⁻
5	.086	.303	.070	.340	.046	.397	5	.083	.295⁻	.068	.325⁺	.045⁻	.388
6	.087	.345⁻	.094	.374	.065⁺	.438	6	.084	.336	.091	.364	.063	.430
7	.134	.385⁺	.103	.413	.086	.477	7	.129	.376	.100	.403	.083	.469
8	.135⁻	.425⁻	.136	.451	.108	.523	8	.130	.416	.131	.440	.104	.505⁺
9	.189	.463	.166	.500	.132	.562	9	.182	.455⁺	.163	.476	.127	.538
10	.190	.500	.184	.549	.157	.603	10	.183	.492	.175⁺	.524	.151	.570
11	.225⁻	.537	.211	.587	.165⁺	.646	11	.219	.524	.205⁺	.560	.152	.612
12	.276	.575⁺	.247	.626	.206	.654	12	.265⁻	.554	.236	.597	.198	.655⁺
13	.294	.615⁻	.251	.660	.211	.684	13	.266	.584	.244	.636	.206	.671
14	.303	.655⁺	.299	.661	.260	.737	14	.295⁻	.624	.292	.675⁺	.249	.692
15	.345⁻	.697	.339	.701	.263	.740	15	.336	.664	.324	.676	.256	.744
16	.385⁺	.706	.340	.749	.316	.789	16	.376	.705⁺	.325⁻	.708	.308	.751
17	.425⁻	.724	.374	.753	.346	.794	17	.416	.734	.364	.756	.329	.794
18	.463	.775⁺	.413	.789	.354	.835⁻	18	.446	.735⁺	.403	.764	.345⁻	.802
19	.500	.810	.451	.816	.397	.843	19	.476	.781	.440	.795⁻	.388	.848
20	.537	.811	.500	.834	.438	.868	20	.508	.817	.476	.825⁻	.430	.849
21	.575⁺	.865⁺	.549	.864	.477	.892	21	.545⁻	.818	.524	.837	.462	.873
22	.615⁻	.866	.587	.897	.523	.914	22	.584	.870	.560	.869	.495⁻	.896
23	.655⁺	.913	.626	.906	.562	.935⁻	23	.624	.871	.597	.900	.531	.917
24	.697	.914	.660	.930	.603	.954	24	.664	.916	.636	.909	.570	.937
25	.721	.938	.701	.951	.646	.970	25	.705⁺	.917	.675⁺	.932	.612	.955⁺
26	.775⁺	.961	.749	.971	.684	.985⁻	26	.734	.941	.708	.953	.655⁺	.972
27	.810	.982	.789	.988	.737	.995⁻	27	.781	.963	.756	.972	.690	.985⁺
28	.865⁺	.996	.834	.998	.789	1.000	28	.817	.982	.795⁻	.988	.744	.995⁻
29	.913	1	.897	1	.840	1	29	.870	.996	.837	.998	.794	1.000
							30	.916	1	.900	1	.848	1

TABLE 21 (Contd.)

(b) ONE-SIDED LIMITS*

If the observed proportion is r/n, enter the table with n and r for an upper one-sided limit. For a lower one-sided limit, enter the table with n and $n - r$ and subtract the table entry from 1.

r	90%	95%	99%	r	90%	95%	99%	r	90%	95%	99%
	$n = 2$				$n = 3$				$n = 4$		
0	.684	.776	.900	0	.536	.632	.785-	0	.438	.527	.684
1	.949	.975-	.995-	1	.804	.865-	.941	1	.680	.751	.859
				2	.965+	.983	.997	2	.857	.902	.958
								3	.974	.987	.997

r	90%	95%	99%	r	90%	95%	99%	r	90%	95%	99%
	$n = 5$				$n = 6$				$n = 7$		
0	.369	.451	.602	0	.319	.393	.536	0	.280	.348	.482
1	.584	.657	.778	1	.510	.582	.706	1	.453	.521	.643
2	.753	.811	.894	2	.667	.729	.827	2	.596	.659	.764
3	.888	.924	.967	3	.799	.847	.915+	3	.721	.775-	.858
4	.979	.990	.998	4	.907	.937	.973	4	.830	.871	.929
				5	.983	.991	.998	5	.921	.947	.977
								6	.985+	.993	.999

r	90%	95%	99%	r	90%	95%	99%	r	90%	95%	99%
	$n = 8$				$n = 9$				$n = 10$		
0	.250	.312	.438	0	.226	.283	.401	0	.206	.259	.369
1	.406	.471	.590	1	.368	.429	.544	1	.337	.394	.504
2	.538	.600	.707	2	.490	.550	.656	2	.450	.507	.612
3	.655+	.711	.802	3	.599	.655+	.750	3	.552	.607	.703
4	.760	.807	.879	4	.699	.749	.829	4	.646	.696	.782
5	.853	.889	.939	5	.790	.831	.895-	5	.733	.778	.850
6	.931	.954	.980	6	.871	.902	.947	6	.812	.850	.907
7	.987	.994	.999	7	.939	.959	.983	7	.884	.913	.952
				8	.988	.994	.999	8	.945+	.963	.984
								9	.990	.995-	.999

r	90%	95%	99%	r	90%	95%	99%	r	90%	95%	99%
	$n = 11$				$n = 12$				$n = 13$		
0	.189	.238	.342	0	.175-	.221	.319	0	.162	.206	.298
1	.310	.364	.470	1	.287	.339	.440	1	.268	.316	.413
2	.415+	.470	.572	2	.386	.438	.537	2	.360	.410	.506
3	.511	.564	.660	3	.475+	.527	.622	3	.444	.495-	.588
4	.599	.650	.738	4	.559	.609	.698	4	.523	.573	.661
5	.682	.729	.806	5	.638	.685-	.765+	5	.598	.645+	.727
6	.759	.800	.866	6	.712	.755-	.825+	6	.669	.713	.787
7	.831	.865-	.916	7	.781	.819	.879	7	.736	.776	.841
8	.895+	.921	.957	8	.846	.877	.924	8	.799	.834	.889
9	.951	.967	.986	9	.904	.928	.961	9	.858	.887	.931
10	.990	.995+	.999	10	.955-	.970	.987	10	.912	.934	.964
				11	.991	.996	.999	11	.958	.972	.988
								12	.992	.996	.999

r	90%	95%	99%	r	90%	95%	99%	r	90%	95%	99%
	$n = 14$				$n = 15$				$n = 16$		
0	.152	.193	.280	0	.142	.181	.264	0	.134	.171	.250
1	.251	.297	.389	1	.236	.279	.368	1	.222	.264	.349
2	.337	.385+	.478	2	.317	.363	.453	2	.300	.344	.430
3	.417	.466	.557	3	.393	.440	.529	3	.371	.417	.503
4	.492	.540	.627	4	.464	.511	.597	4	.439	.484	.569
5	.563	.610	.692	5	.532	.577	.660	5	.504	.548	.630

* Use explained in Sec. 2.3.3. See Sec. 2.3.3 for sample size $n > 30$. This table was computed by Robert S. Gardner, NOTS.

TABLE 21 (Contd.)

(b) One-Sided Limits (Contd.)

r	90%	95%	99%	r	90%	95%	99%	r	90%	95%	99%
	n = 14				n = 15				n = 16		
6	.631	.675⁻	.751	6	.596	.640	.718	6	.565⁺	.609	.687
7	.695⁺	.736	.805⁺	7	.658	.700	.771	7	.625⁻	.667	.739
8	.757	.794	.854	8	.718	.756	.821	8	.682	.721	.788
9	.815⁻	.847	.898	9	.774	.809	.865⁺	9	.737	.773	.834
10	.869	.896	.936	10	.828	.858	.906	10	.790	.822	.875⁻
11	.919	.939	.967	11	.878	.903	.941	11	.839	.868	.912
12	.961	.974	.989	12	.924	.943	.969	12	.886	.910	.945⁻
13	.993	.996	.999	13	.964	.976	.990	13	.929	.947	.971
				14	.993	.997	.999	14	.966	.977	.990
								15	.993	.997	.999
	n = 17				n = 18				n = 19		
0	.127	.162	.237	0	.120	.153	.226	0	.114	.146	.215⁺
1	.210	.250	.332	1	.199	.238	.316	1	.190	.226	.302
2	.284	.326	.410	2	.269	.310	.391	2	.257	.296	.374
3	.352	.396	.480	3	.334	.377	.458	3	.319	.359	.439
4	.416	.461	.543	4	.396	.439	.520	4	.378	.419	.498
5	.478	.522	.603	5	.455⁺	.498	.577	5	.434	.476	.554
6	.537	.580	.658	6	.512	.554	.631	6	.489	.530	.606
7	.594	.636	.709	7	.567	.608	.681	7	.541	.582	.655⁺
8	.650	.689	.758	8	.620	.659	.729	8	.592	.632	.702
9	.703	.740	.803	9	.671	.709	.774	9	.642	.680	.746
10	.754	.788	.845⁻	10	.721	.756	.816	10	.690	.726	.788
11	.803	.834	.883	11	.769	.801	.855⁻	11	.737	.770	.827
12	.849	.876	.918	12	.815⁻	.844	.890	12	.782	.812	.863
13	.893	.915⁺	.948	13	.858	.884	.923	13	.825⁻	.853	.897
14	.933	.950	.973	14	.899	.920	.951	14	.866	.890	.927
15	.968	.979	.991	15	.937	.953	.975⁻	15	.905⁻	.925⁻	.954
16	.994	.997	.999	16	.970	.980	.992	16	.941	.956	.976
				17	.994	.997	.999	17	.972	.981	.992
								18	.994	.997	.999
	n = 20				n = 21				n = 22		
0	.109	.139	.206	0	.104	.133	.197	0	.099	.127	.189
1	.181	.216	.289	1	.173	.207	.277	1	.166	.198	.266
2	.245⁻	.283	.358	2	.234	.271	.344	2	.224	.259	.330
3	.304	.344	.421	3	.291	.329	.404	3	.279	.316	.389
4	.361	.401	.478	4	.345⁺	.384	.460	4	.331	.369	.443
5	.415⁻	.456	.532	5	.397	.437	.512	5	.381	.420	.493
6	.467	.508	.583	6	.448	.487	.561	6	.430	.468	.541
7	.518	.558	.631	7	.497	.536	.608	7	.477	.515⁺	.587
8	.567	.606	.677	8	.544	.583	.653	8	.523	.561	.630
9	.615⁺	.653	.720	9	.590	.628	.695⁺	9	.568	.605⁻	.672
10	.662	.698	.761	10	.636	.672	.736	10	.611	.647	.712
11	.707	.741	.800	11	.679	.714	.774	11	.654	.689	.750
12	.751	.783	.837	12	.722	.755⁺	.811	12	.695⁺	.729	.786
13	.793	.823	.871	13	.764	.794	.845⁺	13	.736	.767	.821
14	.834	.860	.902	14	.804	.832	.878	14	.775⁺	.804	.853
15	.873	.896	.931	15	.842	.868	.908	15	.813	.840	.884
16	.910	.929	.956	16	.879	.901	.935⁻	16	.850	.874	.912
17	.944	.958	.977	17	.914	.932	.959	17	.885⁺	.906	.938
18	.973	.982	.992	18	.946	.960	.978	18	.918	.935⁺	.961
19	.995⁻	.997	.999	19	.974	.983	.993	19	.949	.962	.979
				20	.995⁻	.998	1.000	20	.976	.984	.993
								21	.995⁺	.998	1.000

TABLE 21 (Contd.)

(b) ONE-SIDED LIMITS (Contd.)

r	90%	95%	99%	r	90%	95%	99%	r	90%	95%	99%
	n = 23				n = 24				n = 25		
0	.095+	.122	.181	0	.091	.117	.175−	0	.088	.113	.168
1	.159	.190	.256	1	.153	.183	.246	1	.147	.176	.237
2	.215+	.249	.318	2	.207	.240	.307	2	.199	.231	.296
3	.268	.304	.374	3	.258	.292	.361	3	.248	.282	.349
4	.318	.355−	.427	4	.306	.342	.412	4	.295−	.330	.398
5	.366	.404	.476	5	.352	.389	.460	5	.340	.375+	.444
6	.413	.451	.522	6	.398	.435−	.505−	6	.383	.420	.488
7	.459	.496	.567	7	.442	.479	.548	7	.426	.462	.531
8	.503	.540	.609	8	.484	.521	.590	8	.467	.504	.571
9	.546	.583	.650	9	.526	.563	.630	9	.508	.544	.610
10	.589	.625−	.689	10	.567	.603	.668	10	.548	.583	.648
11	.630	.665−	.727	11	.608	.642	.705−	11	.587	.621	.684
12	.670	.704	.763	12	.647	.681	.740	12	.625−	.659	.719
13	.710	.742	.797	13	.685+	.718	.774	13	.662	.695−	.752
14	.748	.778	.829	14	.723	.754	.806	14	.699	.730	.784
15	.786	.814	.860	15	.759	.788	.837	15	.735−	.764	.815+
16	.822	.848	.889	16	.795+	.822	.867	16	.770	.798	.845+
17	.857	.880	.916	17	.830	.854	.894	17	.804	.830	.873
18	.890	.910	.941	18	.863	.885+	.920	18	.837	.861	.899
19	.922	.938	.962	19	.895+	.914	.943	19	.869	.890	.923
20	.951	.963	.980	20	.925+	.941	.964	20	.899	.918	.946
21	.977	.984	.993	21	.953	.965+	.981	21	.928	.943	.966
22	.995+	.998	1.000	22	.978	.985−	.994	22	.955+	.966	.982
				23	.996	.998	1.000	23	.979	.986	.994
								24	.996	.998	1.000
	n = 26				n = 27				n = 28		
0	.085−	.109	.162	0	.082	.105+	.157	0	.079	.101	.152
1	.142	.170	.229	1	.137	.164	.222	1	.132	.159	.215−
2	.192	.223	.286	2	.185+	.215+	.277	2	.179	.208	.268
3	.239	.272	.337	3	.231	.263	.326	3	.223	.254	.316
4	.284	.318	.385−	4	.275−	.308	.373	4	.265+	.298	.361
5	.328	.363	.430	5	.317	.351	.417	5	.306	.339	.404
6	.370	.405+	.473	6	.358	.392	.458	6	.346	.380	.445−
7	.411	.447	.514	7	.397	.432	.498	7	.385−	.419	.484
8	.451	.487	.554	8	.436	.471	.537	8	.422	.457	.521
9	.491	.526	.592	9	.475−	.509	.574	9	.459	.494	.558
10	.529	.564	.628	10	.512	.547	.610	10	.496	.530	.593
11	.567	.602	.664	11	.549	.583	.645+	11	.532	.565+	.627
12	.604	.638	.698	12	.585−	.618	.679	12	.567	.600	.660
13	.641	.673	.731	13	.620	.653	.711	13	.601	.634	.692
14	.676	.708	.763	14	.655+	.687	.743	14	.635+	.667	.723
15	.711	.742	.794	15	.689	.720	.773	15	.669	.699	.753
16	.746	.774	.823	16	.723	.752	.802	16	.701	.731	.782
17	.779	.806	.851	17	.756	.783	.831	17	.733	.762	.810
18	.812	.837	.878	18	.788	.814	.857	18	.765−	.792	.837
19	.843	.866	.903	19	.819	.843	.883	19	.796	.821	.863
20	.874	.894	.927	20	.849	.871	.907	20	.826	.849	.888
21	.903	.921	.948	21	.879	.899	.930	21	.855+	.876	.911
22	.931	.946	.967	22	.907	.924	.950	22	.883	.902	.932
23	.957	.968	.983	23	.934	.948	.968	23	.911	.927	.952
24	.979	.986	.994	24	.958	.969	.983	24	.936	.950	.969
25	.996	.998	1.000	25	.980	.987	.994	25	.960	.970	.984
				26	.996	.998	1.000	26	.981	.987	.995−
								27	.996	.998	1.000

TABLE 21 (Contd.)

(*b*) ONE-SIDED LIMITS (Contd.)

r	90%	95%	99%	r	90%	95%	99%	r	90%	95%	99%
		$n = 29$				$n = 30$					
0	.076	.098	.147	0	.074	.095+	.142				
1	.128	.153	.208	1	.124	.149	.202				
2	.173	.202	.260	2	.168	.195+	.252				
3	.216	.246	.307	3	.209	.239	.298				
4	.257	.288	.350	4	.249	.280	.340				
5	.297	.329	.392	5	.287	.319	.381				
6	.335-	.368	.432	6	.325-	.357	.420				
7	.372	.406	.470	7	.361	.394	.457				
8	.409	.443	.507	8	.397	.430	.493				
9	.445+	.479	.542	9	.432	.465+	.527				
10	.481	.514	.577	10	.466	.499	.561				
11	.515+	.549	.610	11	.500	.533	.594				
12	.550	.583	.643	12	.533	.566	.626				
13	.583	.616	.674	13	.566	.598	.657				
14	.616	.648	.705-	14	.599	.630	.687				
15	.649	.680	.734	15	.630	.661	.716				
16	.681	.711	.763	16	.662	.692	.744				
17	.712	.741	.791	17	.692	.721	.772				
18	.743	.771	.818	18	.723	.750	.799				
19	.774	.800	.843	19	.752	.779	.824				
20	.803	.828	.868	20	.782	.807	.849				
21	.832	.855-	.892	21	.810	.834	.873				
22	.860	.881	.914	22	.838	.860	.896				
23	.888	.906	.935-	23	.865+	.885+	.917				
24	.914	.930	.954	24	.891	.909	.937				
25	.938	.951	.970	25	.917	.932	.955+				
26	.961	.971	.985-	26	.941	.953	.972				
27	.982	.988	.995-	27	.963	.972	.985+				
28	.996	.998	1.000	28	.982	.988	.995-				
				29	.996	.998	1.000				

APPENDIX CHARTS

CHART I. Rectangular Normal Probability Chart *

$P = 0.001$ per rectangle. One quadrant of the chart is shown.

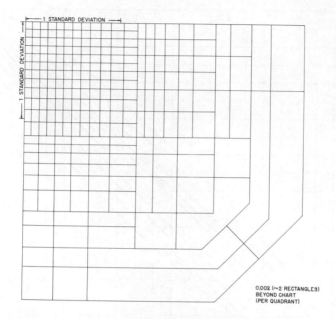

* Use explained in Sec. 1.2.1. This chart was constructed by A. D. Sprague, Bureau of Ordnance, and is reproduced with the permission of R. S. Burington and A. D. Sprague.

270

CHART II. 90% Confidence Belts for Proportions *

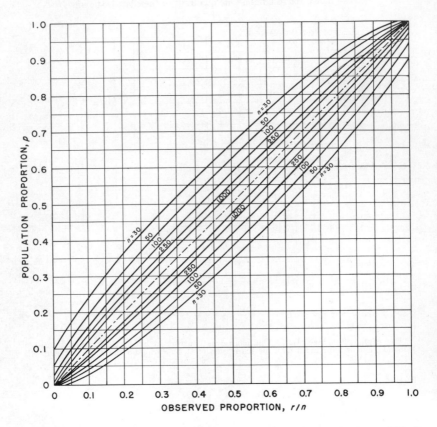

* Use explained in Sec. 2.3.3a. Table 21 should be used for sample size $n \leq 30$. It gives the interval explicitly for each sample size. This chart was constructed by E. L. Crow, NOTS, following the method used by Clopper and Pearson in the construction of Charts III and IV.

CHART III. 95% Confidence Belts for Proportions *

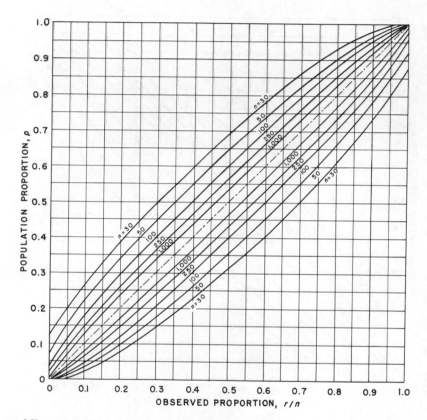

OBSERVED PROPORTION, r/n

° Use explained in Sec. 2.3.3a. Table 21 should be used for sample size $n \leq 30$. It gives the interval explicitly for each sample size. This chart was taken from C. J. Clopper and E. S. Pearson, "The Use of Confidence or Fiducial Limits Illustrated in the Case of the Binomial," BIOMETRIKA, Vol. 26 (1934), pp. 404–13.

272

CHART IV. 99% CONFIDENCE BELTS FOR PROPORTIONS *

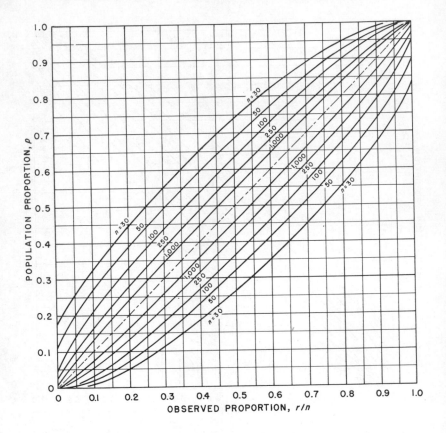

° Use explained in Sec. 2.3.3a. Table 21 should be used for sample size $n \leq 30$. It gives the interval explicitly for each sample size. This chart was taken from C. J. Clopper and E. S. Pearson, "The Use of Confidence or Fiducial Limits Illustrated in the Case of the Binomial," BIOMETRIKA, Vol. 26 (1934), pp. 404–13.

CHART V. OC CURVES FOR TESTING THE HYPOTHESIS $\mu = a$ BY THE EQUAL-TAILS NORMAL TEST *

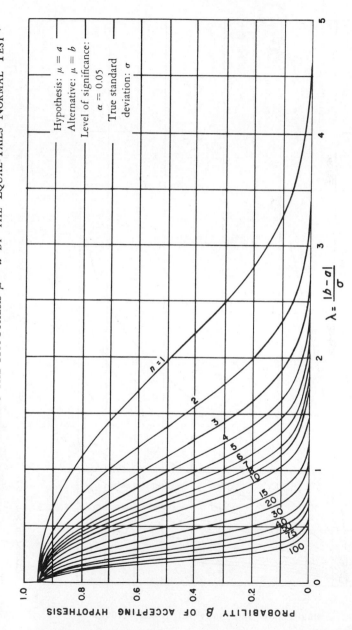

* Use explained in Sec. 2.1.2a. Test statistic $z = \dfrac{\bar{x} - a}{\sigma/\sqrt{n}}$. Adapted, with the permission of the authors and the editor, from C. D. Ferris, F. E. Grubbs, and C. L. Weaver, "Operating Characteristics for the Common Statistical Tests of Significance," ANN MATH STAT, Vol. 17 (1946), p. 190.

CHART VI. OC CURVES FOR TESTING THE HYPOTHESIS $\mu = a$ BY THE EQUAL-TAILS t TEST *

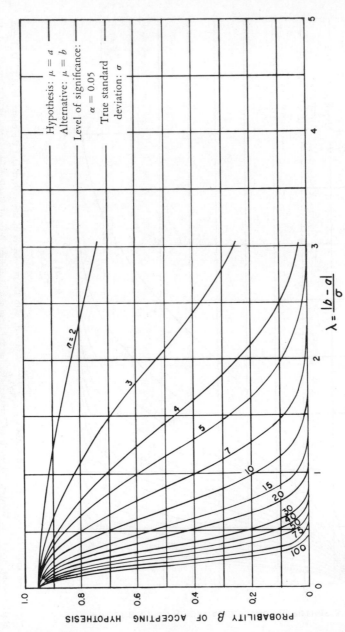

Hypothesis: $\mu = a$
Alternative: $\mu = b$
Level of significance: $\alpha = 0.05$
True standard deviation: σ

$\lambda = \dfrac{|b - a|}{\sigma}$

PROBABILITY β OF ACCEPTING HYPOTHESIS

* Use explained in Sec. 2.2.2a. Test statistic $t = \dfrac{\bar{x} - a}{s/\sqrt{n}}$. Adapted, with the permission of the authors and the editor, from C. D. Ferris, F. E. Grubbs, and C. L. Weaver, "Operating Characteristics for the Common Statistical Tests of Significance," ANN MATH STAT, Vol. 17 (1946), p. 195.

275

CHART VII. OC Curves for Testing the Hypothesis $\sigma = \sigma_0$ Against $\sigma = \sigma_1 > \sigma_0$ by the χ^2 Test *

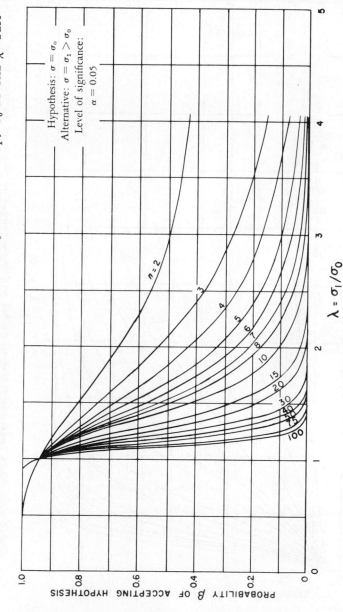

* Use explained in Sec. 3.2.2a. Test statistic $\chi^2 = (n-1)s^2/\sigma_0^2$. Adapted, with the permission of the authors and the editor, from C. D. Ferris, F. E. Grubbs, and C. L. Weaver, "Operating Characteristics for the Common Statistical Tests of Significance," ANN MATH STAT, Vol. 17 (1946), p. 181. Page 183 of this reference gives the OC curves for the alternative $\sigma_1 < \sigma_0$.

276

CHART VIII. OC CURVES FOR TESTING THE HYPOTHESIS $\sigma_1 = \sigma_2$ AGAINST $\sigma_1 > \sigma_2$ BY THE F TEST *

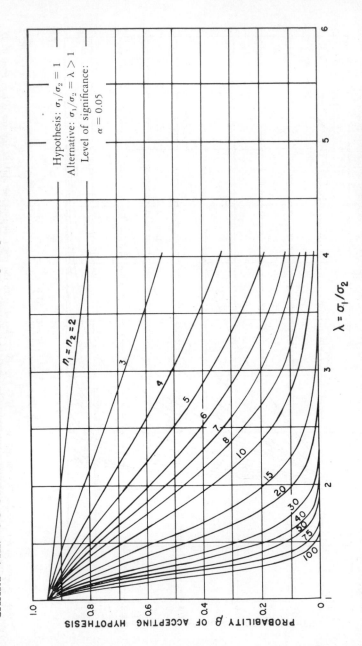

Hypothesis: $\sigma_1/\sigma_2 = 1$
Alternative: $\sigma_1/\sigma_2 = \lambda > 1$
Level of significance: $\alpha = 0.05$

$n_1 = n_2 = 2$

$\lambda = \sigma_1/\sigma_2$

PROBABILITY β OF ACCEPTING HYPOTHESIS

* Use explained in Sec. 3.3.2a. Test Statistic $F = s_1^2/s_2^2$. Adapted, with the permission of the authors and the editor, from C. D. Ferris, F. E. Grubbs, and C. L. Weaver, "Operating Characteristics for the Common Statistical Tests of Significance," ANN MATH STAT, Vol. 17 (1946), p. 185. Pages 186 and 187 of this reference give OC curves for other combinations of (unequal) n_1 and n_2.

CHART IX. NUMBER OF DEGREES OF FREEDOM REQUIRED TO
ESTIMATE THE STANDARD DEVIATION WITHIN $p\%$ OF ITS
TRUE VALUE, WITH PRESCRIBED CONFIDENCE *

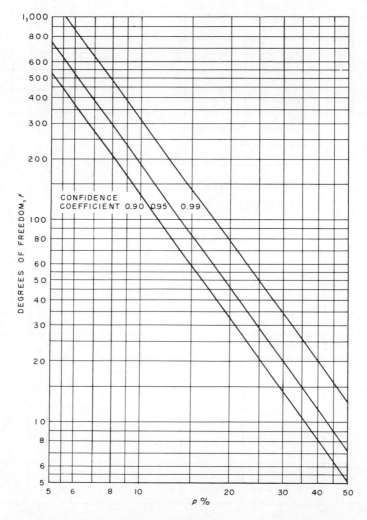

* Use explained in Sec. 3.2.3a; calculate confidence intervals as in Sec. 3.2.3. Adapted, with the permission of the American Statistical Association, from J. A. Greenwood and M. M. Sandomire, "Sample Size Required for Estimating the Standard Deviation as a Percent of Its True Value," AM STAT ASSOC, J, Vol. 45 (1950), p. 258. (The manner of graphing has been changed.)

CHART X. Number of Degrees of Freedom in Each
Sample Required To Estimate σ_1/σ_2 Within $p\%$ of
Its True Value, With Prescribed Confidence *

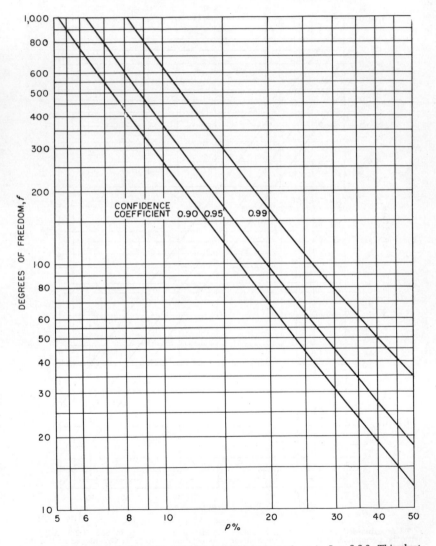

* Use explained in Sec. 3.3.3a; calculate confidence intervals as in Sec. 3.3.3. This chart
was constructed by E. L. Crow and M. W. Maxfield, NOTS.

CHART XI. 95% CONFIDENCE BELTS FOR CORRELATION COEFFICIENT *

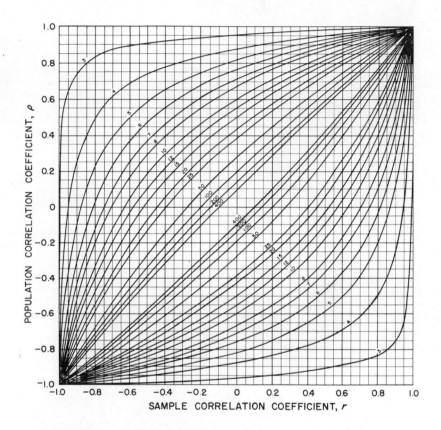

SAMPLE CORRELATION COEFFICIENT, *r*

* Use explained in Sec. 6.1.4b. The numbers on the curves indicate sample size for the case of a two-variable linear regression. Use the curve labeled $n - k + 1$ to find confidence intervals for the partial correlation coefficients in a multiple linear regression with a total of $k + 1$ variables and n observations. Reproduced, with the permission of E. S. Pearson, from F. N. David, *Tables of the Correlation Coefficient*, London, Biometrika Office, 1938, Chart II. Charts for 90, 98, and 99% confidence belts are given in the book.

281

INDEX

A

288

Thompson, C. M., 81, 233, 235, 240
Three-factor analysis of variance, 139
Throwing out readings, 102, *table*, 252
Tightened inspection, 214
Tippett, L. H. C., 208, 248
Tolerance intervals, 18, 104, *table*, 253
Transformation, standardizing, 20, 39
Transformations to obtain linearity, 169, 183, 184
Transformations to obtain normality, 88
Treatments, 111
Trend, described by regression, 147–94
Trend in means, 62, *table*, 250
"True" mean, standard deviation, etc., *see* Population
Truncated sequential plan, 222
Tukey, J. W., 106
Two-factor analysis of variance, 127
Two-variable classification in analysis of variance, 127
Two-variable linear regression, 152
summary of formulas, 164
Types of error, 16, 41
Types of problems in analysis of variance, 118
Types of problems in regression, 148

U

U, upper limit in variables inspection, 209, 212
Unbiased estimate, 12
Unit of inspection, 211
Up-and-down method, 93
Upper limit, U, 209, 212

V

Variable, standardized, 20
Variables testing by control charts or acceptance sampling, 195, 209

Variance, *see* Standard deviation
additivity, 69, 110
analysis of, 109–46
estimated from sample, 12
homogeniety of several variances, 78
of a function of independent variables, 69
population, 12
sample, 12
test of significance, 70
Variation
about the regression line, 156
about the regression plane, 174
Villars, D. S., 146

W

w, range, 13
Wald, Abraham, 226
Wallis, W. A., 107, 194, 205, 226, 253
Weaver, C. L., 273, 274, 275, 276
Wild readings, test for throwing out gross errors, 102, *table*, 252
Wilks, S. S., 35, 65

X

x, mean, 10
$\xi_j(x)$, (xi), orthogonal polynomials, 186

Y

Yates, F., 145, 230, 231, 251
Yates's continuity correction, 99
Youden, W. J., 146, 194

Z

$z = (x - \mu)/\sigma$, standardizing transformation, 20, 39

CONCERNING THE NATURE OF THINGS, Sir William Bragg. Christmas lectures at Royal Society by Nobel laureate, dealing with atoms, gases, liquids, and various types of crystals. No scientific background is needed to understand this remarkably clear introduction to basic processes and aspects of modern science. "More interesting than any bestseller," London Morning Post. 32pp. of photos. 57 figures. xii + 232pp. 5⅜ x 8. **T31 Paperbound $1.35**

THE RISE OF THE NEW PHYSICS, A. d'Abro. Half million word exposition, formerly titled "The Decline of Mechanism," for readers not versed in higher mathematics. Only thorough explanation in everyday language of core of modern mathematical physical theory, treating both classical, modern views. Scientifically impeccable coverage of thought from Newtonian system through theories of Dirac, Heisenberg, Fermi's statistics. Combines history, exposition; broad but unified, detailed view, with constant comparison of classical, modern views. "A must for anyone doing serious study in the physical sciences," J. of the Franklin Inst. "Extraordinary faculty . . . to explain ideas and theories . . . in language of everyday life," Isis. Part I of set: philosophy of science, from practice of Newton, Maxwell, Poincaré, Einstein, etc. Modes of thought, experiment, causality, etc. Part II: 100 pp. on grammar, vocabulary of mathematics, discussions of functions, groups, series, Fourier series, etc. Remainder treats concrete, detailed coverage of both classical, quantum physics: analytic mechanics, Hamilton's principle, electromagnetic waves, thermodynamics, Brownian movement, special relativity, Bohr's atom, de Broglie's wave mechanics, Heisenberg's uncertainty, scores of other important topics. Covers discoveries, theories of d'Alembert, Born, Cantor, Debye, Euler, Foucault, Galois, Gauss, Hadamard, Kelvin, Kepler Laplace, Maxwell, Pauli, Rayleigh Volterra, Weyl, more than 180 others. 97 illustrations. ix + 982pp. 5⅜ x 8.
T3 Vol. 1 Paperbound $2.00
T4 Vol. II Paperbound $2.00

SPINNING TOPS AND GYROSCOPIC MOTION, John Perry. Well-known classic of science still unsurpassed for lucid, accurate, delightful exposition. How quasi-rigidity is induced in flexible, fluid bodies by rapid motions; why gyrostat falls, top rises; nature, effect of internal fluidity on rotating bodies; etc. Appendixes describe practical use of gyroscopes in ships, compasses, monorail transportation. 62 figures. 128pp. 5⅜ x 8.
T416 Paperbound $1.00

FOUNDATIONS OF PHYSICS, R. B. Lindsay, H. Margenau. Excellent bridge between semi-popular and technical writings. Discussion of methods of physical description, construction of theory; valuable to physicist with elementary calculus. Gives meaning to data, tools of modern physics. Contents: symbolism, mathematical equations; space and time; foundations of mechanics; probability; physics, continua; electron theory; relativity; quantum mechanics; causality; etc. "Thorough and yet not overdetailed. Unreservedly recommended," Nature. Unabridged corrected edition. 35 illustrations. xi + 537pp. 5⅜ x 8. **S377 Paperbound $2.45**

FADS AND FALLACIES IN THE NAME OF SCIENCE, Martin Gardner. Formerly entitled "In the Name of Science," the standard account of various cults, quack systems, delusions which have masqueraded as science: hollow earth fanatics, orgone sex energy, dianetics, Atlantis, Forteanism, flying saucers, other recent manifestations. A fair reasoned appraisal of eccentric theory which provides excellent innoculation. "Should be read by everyone, scientist or non-scientist alike," R. T. Birge, Prof. Emeritus of Physics, Univ. of Calif; Former Pres., Amer. Physical Soc. x + 365pp. 5⅜ x 8. **T394 Paperbound $1.50**

ON MATHEMATICS AND MATHEMATICIANS, R. E. Moritz. A 10 year labor of love by discerning, discriminating Prof. Moritz, this collection conveys the full sense of mathematics and personalities of great mathematicians. Anecdotes, aphorisms, reminiscences, philosophies, definitions, speculations, biographical insights, etc. by great mathematicians, writers: Descartes, Mill, Locke, Kant, Coleridge, Whitehead, etc. Glimpses into lives of great mathematicians, from Archimedes to Euler, Gauss, Weierstrass. To mathematicians, a superb browsing-book. To laymen, exciting revelation of fullness of mathematics. Extensive cross index. 410pp. 5⅜ x 8. **T489 Paperbound $1.95**

GUIDE TO THE LITERATURE OF MATHEMATICS AND PHYSICS, N. G. Parke III. Over 5000 entries under approximately 120 major subject headings, of selected most important books, monographs, periodicals, articles in English, plus important works in German, French, Italian, Spanish, Russian (many recently available works). Covers every branch of physics, math, related engineering. Includes author, title, edition, publisher, place, date, number of volumes, number of pages. 40 page introduction on basic problems of research, study provides useful information on organization, use of libraries, psychology of learning, etc. Will save you hours of time. 2nd revised edition. Indices of authors, subjects. 464pp. 5⅜ x 8. **S447 Paperbound $2.49**

THE STRANGE STORY OF THE QUANTUM, An Account for the General Reader of the Growth of Ideas Underlying Our Present Atomic Knowledge, B. Hoffmann. Presents lucidly, expertly, with barest amount of mathematics, problems and theories which led to modern quantum physics. Begins with late 1800's when discrepancies were noticed; with illuminating analogies, examples, goes through concepts of Planck, Einstein, Pauli, Schroedinger, Dirac, Sommerfield, Feynman, etc. New postscript through 1958. "Of the books attempting an account of the history and contents of modern atomic physics which have come to my attention, this is the best," H. Margenau, Yale U., in Amer. J. of Physics. 2nd edition. 32 tables, illustrations. 275pp. 5⅜ x 8. **T518 Paperbound $1.45**

2

Catalogue of Dover
SCIENCE BOOKS

BOOKS THAT EXPLAIN SCIENCE

THE NATURE OF LIGHT AND COLOUR IN THE OPEN AIR, M. Minnaert. Why is falling snow sometimes black? What causes mirages, the fata morgana, multiple suns and moons in the sky; how are shadows formed? Prof. Minnaert of U. of Utrecht answers these and similar questions in optics, light, colour, for non-specialists. Particularly valuable to nature, science students, painters, photographers. "Can best be described in one word—fascinating!" Physics Today. Translated by H. M. Kremer-Priest, K. Jay. 202 illustrations, including 42 photos. xvi + 362pp. 5⅜ x 8.   T196 Paperbound **$1.95**

THE RESTLESS UNIVERSE, Max Born. New enlarged version of this remarkably readable account by a Nobel laureate. Moving from sub-atomic particles to universe, the author explains in very simple terms the latest theories of wave mechanics. Partial contents: air and its relatives, electrons and ions, waves and particles, electronic structure of the atom, nuclear physics. Nearly 1000 illustrations, including 7 animated sequences. 325pp. 6 x 9.   T412 Paperbound **$2.00**

MATTER AND LIGHT, THE NEW PHYSICS, L. de Broglie. Non-technical papers by a Nobel laureate explain electromagnetic theory, relativity, matter, light, radiation, wave mechanics, quantum physics, philosophy of science. Einstein, Planck, Bohr, others explained so easily that no mathematical training is needed for all but 2 of the 21 chapters. "Easy simplicity and lucidity . . . should make this source-book of modern physcis available to a wide public," Saturday Review. Unabridged. 300pp. 5⅜ x 8.   T35 Paperbound **$1.60**

THE COMMON SENSE OF THE EXACT SCIENCES, W. K. Clifford. Introduction by James Newman, edited by Karl Pearson. For 70 years this has been a guide to classical scientific, mathematical thought. Explains with unusual clarity basic concepts such as extension of meaning of symbols, characteristics of surface boundaries, properties of plane figures, vectors, Cartesian method of determining position, etc. Long preface by Bertrand Russell. Bibliography of Clifford. Corrected. 130 diagrams redrawn. 249pp. 5⅜ x 8.
T61 Paperbound **$1.60**

THE EVOLUTION OF SCIENTIFIC THOUGHT FROM NEWTON TO EINSTEIN, A. d'Abro. Einstein's special, general theories of relativity, with historical implications, analyzed in non-technical terms. Excellent accounts of contributions of Newton, Riemann, Weyl, Planck, Eddington, Maxwell, Lorentz, etc., are treated in terms of space, time, equations of electromagnetics, finiteness of universe, methodology of science. "Has become a standard work," Nature. 21 diagrams. 482pp. 5⅜ x 8.   T2 Paperbound **$2.00**

BRIDGES AND THEIR BUILDERS, D. Steinman, S. R. Watson. Engineers, historians, everyone ever fascinated by great spans will find this an endless source of information and interest. Dr. Steinman, recent recipient of Louis Levy Medal, is one of the great bridge architects, engineers of all time. His analysis of great bridges of history is both authoritative and easily followed. Greek, Roman, medieval, oriental bridges; modern works such as Brooklyn Bridge, Golden Gate Bridge, etc. described in terms of history, constructional principles, artistry, function. Most comprehensive, accurate semi-popular history of bridges in print in English. New, greatly revised, enlarged edition. 23 photographs, 26 line drawings. xvii + 401pp. 5⅜ x 8.   T431 Paperbound **$1.95**

HISTORY OF SCIENCE
AND PHILOSOPHY OF SCIENCE

THE VALUE OF SCIENCE, Henri Poincaré. Many of most mature ideas of "last scientific universalist" for both beginning, advanced workers. Nature of scientific truth, whether order is innate in universe or imposed by man, logical thought vs. intuition (relating to Weierstrass, Lie, Riemann, etc), time and space (relativity, psychological time, simultaneity), Herz's concept of force, values within disciplines of Maxwell, Carnot, Mayer, Newton, Lorentz, etc. iii + 147pp. 5⅜ x 8.   S469 Paperbound **$1.35**

PHILOSOPHY AND THE PHYSICISTS, L. S. Stebbing. Philosophical aspects of modern science examined in terms of lively critical attack on ideas of Jeans, Eddington. Tasks of science, causality, determinism, probability, relation of world physics to that of everyday experience, philosophical significance of Planck-Bohr concept of discontinuous energy levels, inferences to be drawn from Uncertainty Principle, implications of "becoming" involved in 2nd law of thermodynamics, other problems posed by discarding of Laplacean determinism. 285pp. 5⅜ x 8.   T480 Paperbound **$1.65**

THE PRINCIPLES OF SCIENCE, A TREATISE ON LOGIC AND THE SCIENTIFIC METHOD, W. S. Jevons. Milestone in development of symbolic logic remains stimulating contribution to investigation of inferential validity in sciences. Treats inductive, deductive logic, theory of number, probability, limits of scientific method; significantly advances Boole's logic, contains detailed introduction to nature and methods of probability in physics, astronomy, everyday affairs, etc. In introduction, Ernest Nagel of Columbia U. says,"[Jevons] continues to be of interest as an attempt to articulate the logic of scientific inquiry." liii + 786pp. 5⅜ x 8.   S446 Paperbound **$2.98**

A HISTORY OF ASTRONOMY FROM THALES TO KEPLER, J. L. E. Dreyer. Only work in English to give complete history of cosmological views from prehistoric times to Kepler. Partial contents: Near Eastern astronomical systems, Early Greeks, Homocentric spheres of Euxodus, Epicycles, Ptolemaic system, Medieval cosmology, Copernicus, Kepler, much more. "Especially useful to teachers and students of the history of science . . . unsurpassed in its field," Isis. Formerly "A History of Planetary Systems from Thales to Kepler." Revised foreword by W. H. Stahl. xvii + 430pp. 5⅜ x 8.   S79 Paperbound **$1.98**

A CONCISE HISTORY OF MATHEMATICS, D. Struik. Lucid study of development of ideas, techniques, from Ancient Near East, Greece, Islamic science, Middle Ages, Renaissance, modern times. Important mathematicians described in detail. Treatment not anecdotal, but analytical development of ideas. Non-technical—no math training needed. "Rich in content, thoughtful in interpretations," U.S. Quarterly Booklist. 60 illustrations including Greek, Egyptian manuscripts, portraits of 31 mathematicians. 2nd edition. xix + 299pp. 5⅜ x 8.   S255 Paperbound **$1.75**

THE PHILOSOPHICAL WRITINGS OF PEIRCE, edited by Justus Buchler. A carefully balanced expositon of Peirce's complete system, written by Peirce himself. It covers such matters as scientific method, pure chance vs. law, symbolic logic, theory of signs, pragmatism, experiment, and other topics. "Excellent selection . . . gives more than adequate evidence of the range and greatness," Personalist. Formerly entitled "The Philosophy of Peirce." xvi + 368pp.   T217 Paperbound **$1.95**

SCIENCE AND METHOD, Henri Poincaré. Procedure of scientific discovery, methodology, experiment, idea-germination—processes by which discoveries come into being. Most significant and interesting aspects of development, application of ideas. Chapters cover selection of facts, chance, mathematical reasoning, mathematics and logic; Whitehead, Russell, Cantor, the new mechanics, etc. 288pp. 5⅜ x 8.   S222 Paperbound **$1.35**

SCIENCE AND HYPOTHESIS, Henri Poincaré. Creative psychology in science. How such concepts as number, magnitude, space, force, classical mechanics developed, how modern scientist uses them in his thought. Hypothesis in physics, theories of modern physics. Introduction by Sir James Larmor. "Few mathematicians have had the breadth of vision of Poincaré, and none is his superior in the gift of clear exposition," E. T. Bell. 272pp. 5⅜ x 8.   S221 Paperbound **$1.35**

ESSAYS IN EXPERIMENTAL LOGIC, John Dewey. Stimulating series of essays by one of most influential minds in American philosophy presents some of his most mature thoughts on wide range of subjects. Partial contents: Relationship between inquiry and experience; dependence of knowledge upon thought; character logic; judgments of practice, data, and meanings; stimuli of thought, etc. viii + 444pp. 5⅜ x 8.   T73 Paperbound **$1.95**

WHAT IS SCIENCE, Norman Campbell. Excellent introduction explains scientific method, role of mathematics, types of scientific laws. Contents: 2 aspects of science, science and nature, laws of chance, discovery of laws, explanation of laws, measurement and numerical laws, applications of science. 192pp. 5⅜ x 8.   S43 Paperbound **$1.25**

FROM EUCLID TO EDDINGTON: A STUDY OF THE CONCEPTIONS OF THE EXTERNAL WORLD, Sir Edmund Whittaker. Foremost British scientist traces development of theories of natural philosophy from western rediscovery of Euclid to Eddington, Einstein, Dirac, etc. 5 major divisions: Space, Time and Movement; Concepts of Classical Physics; Concepts of Quantum Mechanics; Eddington Universe. Contrasts inadequacy of classical physics to understand physical world with present day attempts of relativity, non-Euclidean geometry, space curvature, etc. 212pp. 5⅜ x 8. T491 Paperbound **$1.35**

THE ANALYSIS OF MATTER, Bertrand Russell. How do our senses accord with the new physics? This volume covers such topics as logical analysis of physics, prerelativity physics, causality, scientific inference, physics and perception, special and general relativity, Weyl's theory, tensors, invariants and their physical interpretation, periodicity and qualitative series. "The most thorough treatment of the subject that has yet been published," The Nation. Introduction by L. E. Denonn. 422pp. 5⅜ x 8. T231 Paperbound **$1.95**

LANGUAGE, TRUTH, AND LOGIC, A. Ayer. A clear introduction to the Vienna and Cambridge schools of Logical Positivism. Specific tests to evaluate validity of ideas, etc. Contents: function of philosophy, elimination of metaphysics, nature of analysis, a priori, truth and probability, etc. 10th printing. "I should like to have written it myself," Bertrand Russell. 160pp. 5⅜ x 8. T10 Paperbound **$1.25**

THE PSYCHOLOGY OF INVENTION IN THE MATHEMATICAL FIELD, J. Hadamard. Where do ideas come from? What role does the unconscious play? Are ideas best developed by mathematical reasoning, word reasoning, visualization? What are the methods used by Einstein, Poincaré, Galton, Riemann? How can these techniques be applied by others? One of the world's leading mathematicians discusses these and other questions. xiii + 145pp. 5⅜ x 8. T107 Paperbound **$1.25**

GUIDE TO PHILOSOPHY, C. E. M. Joad. By one of the ablest expositors of all time, this is not simply a history or a typological survey, but an examination of central problems in terms of answers afforded by the greatest thinkers: Plato, Aristotle, Scholastics, Leibniz, Kant, Whitehead, Russell, and many others. Especially valuable to persons in the physical sciences; over 100 pages devoted to Jeans, Eddington, and others, the philosophy of modern physics, scientific materialism, pragmatism, etc. Classified bibliography. 592pp. 5⅜ x 8. T50 Paperbound **$2.00**

SUBSTANCE AND FUNCTION, and EINSTEIN'S THEORY OF RELATIVITY, Ernst Cassirer. Two books bound as one. Cassirer establishes a philosophy of the exact sciences that takes into consideration new developments in mathematics, shows historical connections. Partial contents: Aristotelian logic, Mill's analysis, Helmholtz and Kronecker, Russell and cardinal numbers, Euclidean vs. non-Euclidean geometry, Einstein's relativity. Bibliography. Index. xxi + 464pp. 5⅜ x 8. T50 Paperbound **$2.00**

FOUNDATIONS OF GEOMETRY, Bertrand Russell. Nobel laureate analyzes basic problems in the overlap area between mathematics and philosophy: the nature of geometrical knowledge, the nature of geometry, and the applications of geometry to space. Covers history of non-Euclidean geometry, philosophic interpretations of geometry, especially Kant, projective and metrical geometry. Most interesting as the solution offered in 1897 by a great mind to a problem still current. New introduction by Prof. Morris Kline, N.Y. University. "Admirably clear, precise, and elegantly reasoned analysis," International Math. News. xii + 201pp. 5⅜ x 8. S233 Paperbound **$1.60**

THE NATURE OF PHYSICAL THEORY, P. W. Bridgman. How modern physics looks to a highly unorthodox physicist—a Nobel laureate. Pointing out many absurdities of science, demonstrating inadequacies of various physical theories, weighs and analyzes contributions of Einstein, Bohr, Heisenberg, many others. A non-technical consideration of correlation of science and reality. xi + 138pp. 5⅜ x 8. S33 Paperbound **$1.25**

EXPERIMENT AND THEORY IN PHYSICS, Max Born. A Nobel laureate examines the nature and value of the counterclaims of experiment and theory in physics. Synthetic versus analytical scientific advances are analyzed in works of Einstein, Bohr, Heisenberg, Planck, Eddington, Milne, others, by a fellow scientist. 44pp. 5⅜ x 8. S308 Paperbound **60¢**

A SHORT HISTORY OF ANATOMY AND PHYSIOLOGY FROM THE GREEKS TO HARVEY, Charles Singer. Corrected edition of "The Evolution of Anatomy." Classic traces anatomy, physiology from prescientific times through Greek, Roman periods, dark ages, Renaissance, to beginning of modern concepts. Centers on individuals, movements, that definitely advanced anatomical knowledge. Plato, Diocles, Erasistratus, Galen, da Vinci, etc. Special section on Vesalius. 20 plates. 270 extremely interesting illustrations of ancient, Medieval, enaissance, Oriental origin. xii + 209pp. 5⅜ x 8. T389 Paperbound **$1.75**

SPACE-TIME-MATTER, Hermann Weyl. "The standard treatise on the general theory of relativity," (Nature), by world renowned scientist. Deep, clear discussion of logical coherence of general theory, introducing all needed tools: Maxwell, analytical geometry, non-Euclidean geometry, tensor calculus, etc. Basis is classical space-time, before absorption of relativity. Contents: Euclidean space, mathematical form, metrical continuum, general theory, etc. 15 diagrams. xviii + 330pp. 5⅜ x 8. S267 Paperbound **$1.75**

MATTER AND MOTION, James Clerk Maxwell. Excellent exposition begins with simple particles, proceeds gradually to physical systems beyond complete analysis; motion, force,
properties of centre of mass of material system; work, energy, gravitation, etc. Written
with all Maxwell's original insights and clarity. Notes by E. Larmor. 17 diagrams. 178pp.
5⅜ x 8.   S188 Paperbound **$1.25**

PRINCIPLES OF MECHANICS, Heinrich Hertz. Last work by the great 19th century physicist
is not only a classic, but of great interest in the logic of science. Creating a new system
of mechanics based upon space, time, and mass, it returns to axiomatic analysis, understanding of the formal or structural aspects of science, taking into account logic, observation, a priori elements. Of great historical importance to Poincaré, Carnap, Einstein, Milne.
A 20 page introduction by R. S. Cohen, Wesleyan University, analyzes the implications of
Hertz's thought and the logic of science. 13 page introduction by Helmholtz. xlii + 274pp.
5⅜ x 8.   S316 Clothbound **$3.50**
                                                                   S317 Paperbound **$1.75**

FROM MAGIC TO SCIENCE, Charles Singer. A great historian examines aspects of science
from Roman Empire through Renaissance. Includes perhaps best discussion of early herbals,
penetrating physiological interpretation of "The Visions of Hildegarde of Bingen." Also
examines Arabian, Galenic influences; Pythagoras' sphere, Paracelsus; reawakening of
science under Leonardo da Vinci, Vesalius; Lorica of Gildas the Briton; etc. Frequent
quotations with translations from contemporary manuscripts. Unabridged, corrected edition. 158 unusual illustrations from Classical, Medieval sources. xxvii + 365pp. 5⅜ x 8.
                                                                   T390 Paperbound **$2.00**

A HISTORY OF THE CALCULUS, AND ITS CONCEPTUAL DEVELOPMENT, Carl B. Boyer. Provides
laymen, mathematicians a detailed history of the development of the calculus, from begin-
nings in antiquity to final elaboration as mathematical abstraction. Gives a sense of
mathematics not as technique, but as habit of mind, in progression of ideas of Zeno, Plato,
Pythagoras, Eudoxus, Arabic and Scholastic mathematicians, Newton, Leibniz, Taylor, Descartes, Euler, Lagrange, Cantor, Weierstrass, and others. This first comprehensive, critical
history of the calculus was originally entitled "The Concepts of the Calculus." Foreword
by R. Courant. 22 figures. 25 page bibliography. v + 364pp. 5⅜ x 8.
                                                                   S509 Paperbound **$2.00**

**A DIDEROT PICTORIAL ENCYCLOPEDIA OF TRADES AND INDUSTRY, Manufacturing and the
Technical Arts in Plates Selected from "L'Encyclopédie ou Dictionnaire Raisonné des
Sciences, des Arts, et des Métiers" of Denis Diderot.** Edited with text by C. Gillispie. First
modern selection of plates from high-point of 18th century French engraving. Storehouse
of technological information to historian of arts and science. Over 2,000 illustrations on
485 full page plates, most of them original size, show trades, industries of fascinating
era in such great detail that modern reconstructions might be made of them. Plates teem
with men, women, children performing thousands of operations; show sequence, general
operations, closeups, details of machinery. Illustrates such important, interesting trades,
industries as sowing, harvesting, beekeeping, tobacco processing, fishing, arts of war,
mining, smelting, casting iron, extracting mercury, making gunpowder, cannons, bells,
shoeing horses, tanning, papermaking, printing, dying, over 45 more categories. Professor
Gillispie of Princeton supplies full commentary on all plates, identifies operations, tools,
processes, etc. Material is presented in lively, lucid fashion. Of great interest to all
studying history of science, technology. Heavy library cloth. 920pp. 9 x 12.
                                                                   T421 2 volume set **$18.50**

DE MAGNETE, William Gilbert. Classic work on magnetism, founded new science. Gilbert
was first to use word "electricity," to recognize mass as distinct from weight, to discover
effect of heat on magnetic bodies; invented an electroscope, differentiated between static
electricity and magnetism, conceived of earth as magnet. This lively work, by first great
experimental scientist, is not only a valuable historical landmark, but a delightfully easy
to follow record of a searching, ingenious mind. Translated by P. F. Mottelay. 25 page
biographical memoir. 90 figures. lix + 368pp. 5⅜ x 8.   S470 Paperbound **$2.00**

HISTORY OF MATHEMATICS, D. E. Smith. Most comprehensive, non-technical history of math
in English. Discusses lives and works of over a thousand major, minor figures, with foot-
notes giving technical information outside book's scheme, and indicating disputed matters.
Vol. I: A chronological examination, from primitive concepts through Egypt, Babylonia,
Greece, the Orient, Rome, the Middle Ages, The Renaissance, and to 1900. Vol. II: The
development of ideas in specific fields and problems, up through elementary calculus.
"Marks an epoch . . . will modify the entire teaching of the history of science," George
Sarton. 2 volumes, total of 510 illustrations, 1355pp. 5⅜ x 8. Set boxed in attractive
container.   T429, 430 Paperbound, the set **$5.00**

THE PHILOSOPHY OF SPACE AND TIME, H. Reichenbach. An important landmark in development of empiricist conception of geometry, covering foundations of geometry, time theory,
consequences of Einstein's relativity, including: relations between theory and observations;
coordinate definitions; relations between topological and metrical properties of space;
psychological problem of visual intuition of non-Euclidean structures; many more topics
important to modern science and philosophy. Majority of ideas require only knowledge of
intermediate math. "Still the best book in the field," Rudolf Carnap. Introduction by
R. Carnap. 49 figures. xviii + 296pp. 5⅜ x 8.   S443 Paperbound **$2.00**

FOUNDATIONS OF SCIENCE: THE PHILOSOPHY OF THEORY AND EXPERIMENT, N. Campbell.
A critique of the most fundamental concepts of science, particularly physics. Examines why certain propositions are accepted without question, demarcates science from philosophy, etc. Part I analyzes presuppositions of scientific thought: existence of material world, nature of laws, probability, etc; part 2 covers nature of experiment and applications of mathematics: conditions for measurement, relations between numerical laws and theories, error, etc. An appendix covers problems arising from relativity, force, motion, space, time. A classic in its field. "A real grasp of what science is," Higher Educational Journal. xiii + 565pp. 5⅝ x 8⅜.   S372 Paperbound **$2.95**

THE STUDY OF THE HISTORY OF MATHEMATICS and **THE STUDY OF THE HISTORY OF SCIENCE, G. Sarton.** Excellent introductions, orientation, for beginning or mature worker. Describes duty of mathematical historian, incessant efforts and genius of previous generations. Explains how today's discipline differs from previous methods. 200 item bibliography with critical evaluations, best available biographies of modern mathematicians, best treatises on historical methods is especially valuable. 10 illustrations. 2 volumes bound as one. 113pp. + 75pp. 5⅜ x 8.   T240 Paperbound **$1.25**

MATHEMATICAL PUZZLES

MATHEMATICAL PUZZLES OF SAM LOYD, selected and edited by **Martin Gardner.** 117 choice puzzles by greatest American puzzle creator and innovator, from his famous "Cyclopedia of Puzzles." All unique style, historical flavor of originals. Based on arithmetic, algebra, probability, game theory, route tracing, topology, sliding block, operations research, geometrical dissection. Includes famous "14-15" puzzle which was national craze, "Horse of a Different Color" which sold millions of copies. 120 line drawings, diagrams. Solutions. xx + 167pp. 5⅜ x 8.   T498 Paperbound **$1.00**

SYMBOLIC LOGIC and THE GAME OF LOGIC, Lewis Carroll. "Symbolic Logic" is not concerned with modern symbolic logic, but is instead a collection of over 380 problems posed with charm and imagination, using the syllogism, and a fascinating diagrammatic method of drawing conclusions. In "The Game of Logic" Carroll's whimsical imagination devises a logical game played with 2 diagrams and counters (included) to manipulate hundreds of tricky syllogisms. The final section, "Hit or Miss" is a lagniappe of 101 additional puzzles in the delightful Carroll manner. Until this reprint edition, both of these books were rarities costing up to $15 each. Symbolic Logic: Index. xxxi + 199pp. The Game of Logic: 96pp. 2 vols. bound as one. 5⅜ x 8.   T492 Paperbound **$1.50**

PILLOW PROBLEMS and A TANGLED TALE, Lewis Carroll. One of the rarest of all Carroll's works, "Pillow Problems" contains 72 original math puzzles, all typically ingenious. Particularly fascinating are Carroll's answers which remain exactly as he thought them out, reflecting his actual mental process. The problems in "A Tangled Tale" are in story form, originally appearing as a monthly magazine serial. Carroll not only gives the solutions, but uses answers sent in by readers to discuss wrong approaches and misleading paths, and grades them for insight. Both of these books were rarities until this edition, "Pillow Problems" costing up to $25, and "A Tangled Tale" $15. Pillow Problems: Preface and Introduction by Lewis Carroll. xx + 109pp. A Tangled Tale: 6 illustrations. 152pp. Two vols. bound as one. 5⅜ x 8.   T493 Paperbound **$1.50**

NEW WORD PUZZLES, G. L. Kaufman. 100 brand new challenging puzzles on words, combinations, never before published. Most are new types invented by author, for beginners and experts both. Squares of letters follow chess moves to build words; symmetrical designs made of synonyms; rhymed crostics; double word squares; syllable puzzles where you fill in missing syllables instead of missing letter; many other types, all new. Solutions. "Excellent," Recreation. 100 puzzles. 196 figures. vi + 122pp. 5⅜ x 8.
                                                                T344 Paperbound **$1.00**

MATHEMATICAL EXCURSIONS, H. A. Merrill. Fun, recreation, insights into elementary problem solving. Math expert guides you on by-paths not generally travelled in elementary math courses—divide by inspection, Russian peasant multiplication; memory systems for pi; odd, even magic squares; dyadic systems; square roots by geometry; Tchebichev's machine; dozens more. Solutions to more difficult ones. "Brain stirring stuff . . . a classic," Genie. 50 illustrations. 145pp. 5⅜ x 8.   T350 Paperbound **$1.00**

THE BOOK OF MODERN PUZZLES, G. L. Kaufman. Over 150 puzzles, absolutely all new material based on same appeal as crosswords, deduction puzzles, but with different principles, techniques. 2-minute teasers, word labyrinths, design, pattern, logic, observation puzzles, puzzles testing ability to apply general knowledge to peculiar situations, many others. Solutions. 116 illustrations. 192pp. 5⅜ x 8.   T143 Paperbound **$1.00**

MATHEMAGIC, MAGIC PUZZLES, AND GAMES WITH NUMBERS, R. V. Heath. Over 60 puzzles, stunts, on properties of numbers. Easy techniques for multiplying large numbers mentally, identifying unknown numbers, finding date of any day in any year. Includes The Lost Digit, 3 Acrobats, Psychic Bridge, magic squares, triangles, cubes, others not easily found elsewhere. Edited by J. S. Meyer. 76 illustrations. 128pp. 5⅜ x 8.   T110 Paperbound **$1.00**

DOVER SCIENCE BOOKS

PUZZLE QUIZ AND STUNT FUN, J. Meyer. 238 high-priority puzzles, stunts, tricks—math puzzles like The Clever Carpenter, Atom Bomb, Please Help Alice; mysteries, deductions like The Bridge of Sighs, Secret Code; observation puzzlers like The American Flag, Playing Cards, Telephone Dial; over 200 others with magic squares, tongue twisters, puns, anagrams. Solutions. Revised, enlarged edition of "Fun-To-Do." Over 100 illustrations. 238 puzzles, stunts, tricks. 256pp. 5⅜ x 8.   T337 Paperbound **$1.00**

101 PUZZLES IN THOUGHT AND LOGIC, C. R. Wylie, Jr. For readers who enjoy challenge, stimulation of logical puzzles without specialized math or scientific knowledge. Problems entirely new, range from relatively easy to brainteasers for hours of subtle entertainment. Detective puzzles, find the lying fisherman, how a blind man identifies color by logic, many more. Easy-to-understand introduction to logic of puzzle solving and general scientific method. 128pp. 5⅜ x 8.   T367 Paperbound **$1.00**

CRYPTANALYSIS, H. F. Gaines. Standard elementary, intermediate text for serious students. Not just old material, but much not generally known, except to experts. Concealment, Transposition, Substitution ciphers; Vigenere, Kasiski, Playfair, multafid, dozens of other techniques. Formerly "Elementary Cryptanalysis." Appendix with sequence charts, letter frequencies in English, 5 other languages, English word frequencies. Bibliography. 167 codes. New to this edition: solutions to codes. vi + 230pp. 5⅜ x 8⅜.
T97 Paperbound **$1.95**

CRYPTOGRAPY, L. D. Smith. Excellent elementary introduction to enciphering, deciphering secret writing. Explains transposition, substitution ciphers; codes; solutions; geometrical patterns, route transcription, columnar transposition, other methods. Mixed cipher systems; single, polyalphabetical substitutions; mechanical devices; Vigenere; etc. Enciphering Japanese; explanation of Baconian biliteral cipher; frequency tables. Over 150 problems. Bibliography. Index. 164pp. 5⅜ x 8.   T247 Paperbound **$1.00**

MATHEMATICS, MAGIC AND MYSTERY, M. Gardner. Card tricks, metal mathematics, stage mind-reading, other "magic" explained as applications of probability, sets, number theory, etc. Creative examination of laws, applications. Scores of new tricks, insights. 115 sections on cards, dice, coins; vanishing tricks, many others. No sleight of hand—math guarantees success. "Could hardly get more entertainment . . . easy to follow," Mathematics Teacher. 115 illustrations. xii + 174pp. 5⅜ x 8.   T335 Paperbound **$1.00**

AMUSEMENTS IN MATHEMATICS, H. E. Dudeney. Foremost British originator of math puzzles, always witty, intriguing, paradoxical in this classic. One of largest collections. More than 430 puzzles, problems, paradoxes. Mazes, games, problems on number manipulations, unicursal, other route problems, puzzles on measuring, weighing, packing, age, kinship, chessboards, joiners', crossing river, plane figure dissection, many others. Solutions. More than 450 illustrations. viii + 258pp. 5⅜ x 8.   T473 Paperbound **$1.25**

THE CANTERBURY PUZZLES H. E. Dudeney. Chaucer's pilgrims set one another problems in story form. Also Adventures of the Puzzle Club, the Strange Escape of the King's Jester, the Monks of Riddlewell, the Squire's Christmas Puzzle Party, others. All puzzles are original, based on dissecting plane figures, arithmetic, algebra, elementary calculus, other branches of mathematics, and purely logical ingenuity. "The limit of ingenuity and intricacy," The Observer. Over 110 puzzles, full solutions. 150 illustrations. viii + 225 pp. 5⅜ x 8.   T474 Paperbound **$1.25**

MATHEMATICAL PUZZLES FOR BEGINNERS AND ENTHUSIASTS, G. Mott-Smith. 188 puzzles to test mental agility. Inference, interpretation, algebra, dissection of plane figures, geometry, properties of numbers, decimation, permutations, probability, all are in these delightful problems. Includes the Odic Force, How to Draw an Ellipse, Spider's Cousin, more than 180 others. Detailed solutions. Appendix with square roots, triangular numbers, primes, etc. 135 illustrations. 2nd revised edition. 248pp. 5⅜ x 8.   T198 Paperbound **$1.00**

MATHEMATICAL RECREATIONS, M. Kraitchik. Some 250 puzzles, problems, demonstrations of recreation mathematics on relatively advanced level. Unusual historical problems from Greek, Medieval, Arabic, Hindu sources; modern problems on "mathematics without numbers," geometry, topology, arithmetic, etc. Pastimes derived from figurative, Mersenne, Fermat numbers: fairy chess; latruncles: reversi; etc. Full solutions. Excellent insights into special fields of math. "Strongly recommended to all who are interested in the lighter side of mathematics," Mathematical Gaz. 181 illustrations. 330pp. 5⅜ x 8.
T163 Paperbound **$1.75**

FICTION

FLATLAND, E. A. Abbott. A perennially popular science-fiction classic about life in a 2-dimensional world, and the impingement of higher dimensions. Political, satiric, humorous, moral overtones. This land where women are straight lines and the lowest and most dangerous classes are isosceles triangles with 3° vertices conveys brilliantly a feeling for many concepts of modern science. 7th edition. New introduction by Banesh Hoffmann. 128pp. 5⅜ x 8.   T1 Paperbound **$1.00**

SEVEN SCIENCE FICTION NOVELS OF H. G. WELLS. Complete texts, unabridged, of seven of Wells' greatest novels: The War of the Worlds, The Invisible Man, The Island of Dr. Moreau, The Food of the Gods, First Men in the Moon, In the Days of the Comet, The Time Machine. Still considered by many experts to be the best science-fiction ever written, they will offer amusements and instruction to the scientific minded reader. "The great master," Sky and Telescope. 1051pp. 5⅜ x 8. T264 Clothbound **$3.95**

28 SCIENCE FICTION STORIES OF H. G. WELLS. Unabridged! This enormous omnibus contains 2 full length novels—Men Like Gods, Star Begotten—plus 26 short stories of space, time, invention, biology, etc. The Crystal Egg, The Country of the Blind, Empire of the Ants, The Man Who Could Work Miracles, Aepyornis Island, A Story of the Days to Come, and 20 others "A master . . . not surpassed by . . . writers of today," The English Journal. 915pp. 5⅜ x 8. T265 Clothbound **$3.95**

FIVE ADVENTURE NOVELS OF H. RIDER HAGGARD. All the mystery and adventure of darkest Africa captured accurately by a man who lived among Zulus for years, who knew African ethnology, folkways as did few of his contemporaries. They have been regarded as examples of the very best high adventure by such critics as Orwell, Andrew Lang, Kipling. Contents: She, King Solomon's Mines, Allan Quatermain, Allan's Wife, Maiwa's Revenge. "Could spin a yarn so full of suspense and color that you couldn't put the story down," Sat. Review. 821pp. 5⅜ x 8. T108 Clothbound **$3.95**

CHESS AND CHECKERS

LEARN CHESS FROM THE MASTERS, Fred Reinfeld. Easiest, most instructive way to improve your game—play 10 games against such masters as Marshall, Znosko-Borovsky, Bronstein, Najdorf, etc., with each move graded by easy system. Includes ratings for alternate moves possible. Games selected for interest, clarity, easily isolated principles. Covers Ruy Lopez, Dutch Defense, Vienna Game openings; subtle, intricate middle game variations; all-important end game. Full annotations. Formerly "Chess by Yourself." 91 diagrams. viii + 144pp. 5⅜ x 8. T362 Paperbound **$1.00**

REINFELD ON THE END GAME IN CHESS, Fred Reinfeld. Analyzes 62 end games by Alekhine, Flohr, Tarrasch, Morphy, Capablanca, Rubinstein, Lasker, Reshevsky, other masters. Only 1st rate book with extensive coverage of error—tell exactly what is wrong with each move you might have made. Centers around transitions from middle play to end play. King and pawn, minor pieces, queen endings; blockage, weak, passed pawns, etc. "Excellent . . . a boon," Chess Life. Formerly "Practical End Play." 62 figures. vi + 177pp. 5⅜ x 8. T417 Paperbound **$1.25**

HYPERMODERN CHESS as developed in the games of its greatest exponent, ARON NIMZO-VICH, edited by Fred Reinfeld. An intensely original player, analyst, Nimzovich's approaches startled, often angered the chess world. This volume, designed for the average player, shows how his iconoclastic methods won him victories over Alekhine, Lasker, Marshall, Rubinstein, Spielmann, others, and infused new life into the game. Use his methods to startle opponents, invigorate play. "Annotations and introductions to each game . . . are excellent," Times (London). 180 diagrams. viii + 220pp. 5⅜ x 8. T448 Paperbound **$1.35**

THE ADVENTURE OF CHESS, Edward Lasker. Lively reader, by one of America's finest chess masters, including: history of chess, from ancient Indian 4-handed game of Chaturanga to great players of today; such delights and oddities as Maelzel's chess-playing automaton that beat Napoleon 3 times; etc. One of most valuable features is author's personal recollections of men he has played against—Nimzovich, Emanuel Lasker, Capablanca, Alekhine, etc. Discussion of chess-playing machines (newly revised). 5 page chess primer. 11 illustrations. 53 diagrams. 296pp. 5⅜ x 8. S510 Paperbound **$1.45**

THE ART OF CHESS, James Mason. Unabridged reprinting of latest revised edition of most famous general study ever written. Mason, early 20th century master, teaches beginning, intermediate player over 90 openings; middle game, end game, to see more moves ahead, to plan purposefully, attack, sacrifice, defend, exchange, govern general strategy. "Classic . . . one of the clearest and best developed studies," Publishers Weekly. Also included, a complete supplement by F. Reinfeld, "How Do You Play Chess?", invaluable to beginners for its lively question-and-answer method. 448 diagrams. 1947 Reinfeld-Bernstein text. Bibliography. xvi + 340pp. 5⅜ x 8. T463 Paperbound **$1.85**

MORPHY'S GAMES OF CHESS, edited by P. W. Sergeant. Put boldness into your game by flowing brilliant, forceful moves of the greatest chess player of all time. 300 of Morphy's best games, carefully annotated to reveal principles. 54 classics against masters like Anderssen, Harrwitz, Bird, Paulsen, and others. 52 games at odds; 54 blindfold games; plus over 100 others. Follow his interpretation of Dutch Defense, Evans Gambit, Giuoco Piano, Ruy Lopez, many more. Unabridged reissue of latest revised edition. New introduction by F. Reinfeld. Annotations, introduction by Sergeant. 235 diagrams. x + 352pp. 5⅜ x 8. T386 Paperbound **$1.75**

WIN AT CHECKERS, M. Hopper. (Formerly "Checkers.") Former World's Unrestricted Checker Champion discusses principles of game, expert's shots, traps, problems for beginner, standard openings, locating best move, end game, opening "blitzkrieg" moves to draw when behind, etc. Over 100 detailed questions, answers anticipate problems. Appendix. 75 problems with solutions, diagrams. 79 figures. xi + 107pp. 5⅜ x 8.   T363 Paperbound **$1.00**

HOW TO FORCE CHECKMATE, Fred Reinfeld. If you have trouble finishing off your opponent, here is a collection of lightning strokes and combinations from actual tournament play. Starts with 1-move checkmates, works up to 3-move mates. Develops ability to lock ahead, gain new insights into combinations, complex or deceptive positions; ways to estimate weaknesses, strengths of you and your opponent. "A good deal of amusement and instruction," Times, (London). 300 diagrams. Solutions to all positions. Formerly "Challenge to Chess Players." 111pp. 5⅜ x 8.   T417 Paperbound **$1.25**

A TREASURY OF CHESS LORE, edited by Fred Reinfeld. Delightful collection of anecdotes, short stories, aphorisms by, about masters; poems, accounts of games, tournaments, photographs; hundreds of humorous, pithy, satirical, wise, historical episodes, comments, word portraits. Fascinating "must" for chess players; revealing and perhaps seductive to those who wonder what their friends see in game. 49 photographs (14 full page plates). 12 diagrams. xi + 306pp. 5⅜ x 8.   T458 Paperbound **$1.75**

WIN AT CHESS, Fred Reinfeld. 300 practical chess situations, to sharpen your eye, test skill against masters. Start with simple examples, progress at own pace to complexities. This selected series of crucial moments in chess will stimulate imagination, develop stronger, more versatile game. Simple grading system enables you to judge progress. "Extensive use of diagrams is a great attraction," Chess. 300 diagrams. Notes, solutions to every situation. Formerly "Chess Quiz." vi + 120pp. 5⅜ x 8.   T433 Paperbound **$1.00**

MATHEMATICS:
ELEMENTARY TO INTERMEDIATE

HOW TO CALCULATE QUICKLY, H. Sticker. Tried and true method to help mathematics of everyday life. Awakens "number sense"—ability to see relationships between numbers as whole quantities. A serious course of over 9000 problems and their solutions through techniques not taught in schools: left-to-right multiplications, new fast division, etc. 10 minutes a day will double or triple calculation speed. Excellent for scientist at home in higher math, but dissatisfied with speed and accuracy in lower math. 256pp. 5 x 7¼.
Paperbound **$1.00**

FAMOUS PROBLEMS OF ELEMENTARY GEOMETRY, Felix Klein. Expanded version of 1894 Easter lectures at Göttingen. 3 problems of classical geometry: squaring the circle, trisecting angle, doubling cube, considered with full modern implications: transcendental numbers, pi, etc. "A modern classic . . . no knowledge of higher mathematics is required," Scientia. Notes by R. Archibald. 16 figures. xi + 92pp. 5⅜ x 8.   T298 Paperbound **$1.00**

HIGHER MATHEMATICS FOR STUDENTS OF CHEMISTRY AND PHYSICS, J. W. Mellor. Practical, not abstract, building problems out of familiar laboratory material. Covers differential calculus, coordinate, analytical geometry, functions, integral calculus, infinite series, numerical equations, differential equations, Fourier's theorem probability, theory of errors, calculus of variations, determinants. "If the reader is not familiar with this book, it will repay him to examine it," Chem. and Engineering News. 800 problems. 189 figures. xxi + 641pp. 5⅜ x 8.   S193 Paperbound **$2.25**

TRIGONOMETRY REFRESHER FOR TECHNICAL MEN, A. A. Klaf. 913 detailed questions, answers cover most important aspects of plane, spherical trigonometry—particularly useful in clearing up difficulties in special areas. Part I: plane trig, angles, quadrants, functions, graphical representation, interpolation, equations, logs, solution of triangle, use of slide rule, etc. Next 188 pages discuss applications to navigation, surveying, elasticity, architecture, other special fields. Part 3: spherical trig, applications to terrestrial, astronomical problems. Methods of time-saving, simplification of principal angles, make book most useful. 913 questions answered. 1738 problems, answers to odd numbers. 494 figures. 24 pages of formulas, functions. x + 629pp. 5⅜ x 8.   T371 Paperbound **$2.00**

CALCULUS REFRESHER FOR TECHNICAL MEN, A. A. Klaf. 756 questions examine most important aspects of integral, differential calculus. Part I: simple differential calculus, constants, variables, functions, increments, logs, curves, etc. Part 2: fundamental ideas of integrations, inspection, substitution, areas, volumes, mean value, double, triple integration, etc. Practical aspects stressed. 50 pages illustrate applications to specific problems of civil, nautical engineering, electricity, stress, strain, elasticity, similar fields. 756 questions answered. 566 problems, mostly answered. 36pp. of useful constants, formulas. v + 431pp. 5⅜ x 8.   T370 Paperbound **$2.00**

9

CATALOGUE OF

MONOGRAPHS ON TOPICS OF MODERN MATHEMATICS, edited by J. W. A. Young. Advanced mathematics for persons who have forgotten, or not gone beyond, high school algebra. 9 monographs on foundation of geometry, modern pure geometry, non-Euclidean geometry, fundamental propositions of algebra, algebraic equations, functions, calculus, theory of numbers, etc. Each monograph gives proofs of important results, and descriptions of leading methods, to provide wide coverage. "Of high merit," Scientific American. New introduction by Prof. M. Kline, N.Y. Univ. 100 diagrams. xvi + 416pp. 6⅛ x 9¼.
S289 Paperbound **$2.00**

MATHEMATICS IN ACTION, O. G. Sutton. Excellent middle level application of mathematics to study of universe, demonstrates how math is applied to ballistics, theory of computing machines, waves, wave-like phenomena, theory of fluid flow, meteorological problems, statistics, flight, similar phenomena. No knowledge of advanced math required. Differential equations, Fourier series, group concepts, Eigenfunctions, Planck's constant, airfoil theory, and similar topics explained so clearly in everyday language that almost anyone can derive benefit from reading this even if much of high-school math is forgotten. 2nd edition. 88 figures. viii + 236pp. 5⅜ x 8.
T450 Clothbound **$3.50**

ELEMENTARY MATHEMATICS FROM AN ADVANCED STANDPOINT, Felix Klein. Classic text, an outgrowth of Klein's famous integration and survey course at Göttingen. Using one field to interpret, adjust another, it covers basic topics in each area, with extensive analysis. Especially valuable in areas of modern mathematics. "A great mathematician, inspiring teacher, . . . deep insight," Bul., Amer. Math Soc.

Vol. I. ARITHMETIC, ALGEBRA, ANALYSIS. Introduces concept of function immediately, enlivens discussion with graphical, geometric methods. Partial contents: natural numbers, special properties, complex numbers. Real equations with real unknowns, complex quantities. Logarithmic, exponential functions, infinitesimal calculus. Transcendence of e and pi, theory of assemblages. Index. 125 figures. ix + 274pp. 5⅜ x 8.
S151 Paperbound **$1.75**

Vol. II. GEOMETRY. Comprehensive view, accompanies space perception inherent in geometry with analytic formulas which facilitate precise formulation. Partial contents: Simplest geometric manifold; line segments, Grassman determinant principles, classification of configurations of space. Geometric transformations: affine, projective, higher point transformations, theory of the imaginary. Systematic discussion of geometry and its foundations. 141 illustrations. ix + 214pp. 5⅜ x 8.
S151 Paperbound **$1.75**

A TREATISE ON PLANE AND ADVANCED TRIGONOMETRY, E. W. Hobson. Extraordinarily wide coverage, going beyond usual college level, one of few works covering advanced trig in full detail. By a great expositor with unerring anticipation of potentially difficult points. Includes circular functions; expansion of functions of multiple angle; trig tables; relations between sides, angles of triangles; complex numbers; etc. Many problems fully solved. "The best work on the subject," Nature. Formerly entitled "A Treatise on Plane Trigonometry." 689 examples. 66 figures. xvi + 383pp. 5⅜ x 8.
S353 Paperbound **$1.95**

NON-EUCLIDEAN GEOMETRY, Roberto Bonola. The standard coverage of non-Euclidean geometry. Examines from both a historical and mathematical point of view geometries which have arisen from a study of Euclid's 5th postulate on parallel lines. Also included are complete texts, translated, of Bolyai's "Theory of Absolute Space," Lobachevsky's "Theory of Parallels." 180 diagrams. 431pp. 5⅜ x 8.
S27 Paperbound **$1.95**

GEOMETRY OF FOUR DIMENSIONS, H. P. Manning. Unique in English as a clear, concise introduction. Treatment is synthetic, mostly Euclidean, though in hyperplanes and hyperspheres at infinity, non-Euclidean geometry is used. Historical introduction. Foundations of 4-dimensional geometry. Perpendicularity, simple angles. Angles of planes, higher order. Symmetry, order, motion; hyperpyramids, hypercones, hyperspheres; figures with parallel elements; volume, hypervolume in space; regular polyhedroids. Glossary. 78 figures. ix + 348pp. 5⅜ x 8.
S182 Paperbound **$1.95**

MATHEMATICS: INTERMEDIATE TO ADVANCED

GEOMETRY (EUCLIDEAN AND NON-EUCLIDEAN)

THE GEOMETRY OF RENÉ DESCARTES. With this book, Descartes founded analytical geometry. Original French text, with Descartes's own diagrams, and excellent Smith-Latham translation. Contains: Problems the Construction of Which Requires only Straight Lines and Circles; On the Nature of Curved Lines; On the Construction of Solid or Supersolid Problems. Diagrams. 258pp. 5⅜ x 8.
S68 Paperbound **$1.50**

THE WORKS OF ARCHIMEDES, edited by T. L. Heath. All the known works of the great Greek mathematician, including the recently discovered Method of Archimedes. Contains: On Sphere and Cylinder, Measurement of a Circle, Spirals, Conoids, Spheroids, etc. Definitive edition of greatest mathematical intellect of ancient world. 186 page study by Heath discusses Archimedes and history of Greek mathematics. 563pp. 5⅜ x 8. S9 Paperbound **$2.00**

COLLECTED WORKS OF BERNARD RIEMANN. Important sourcebook, first to contain complete text of 1892 "Werke" and the 1902 supplement, unabridged. 31 monographs, 3 complete lecture courses, 15 miscellaneous papers which have been of enormous importance in relativity, topology, theory of complex variables, other areas of mathematics. Edited by R. Dedekind, H. Weber, M. Noether, W. Wirtinger. German text; English introduction by Hans Lewy. 690pp. 5⅜ x 8. S226 Paperbound **$2.85**

THE THIRTEEN BOOKS OF EUCLID'S ELEMENTS, edited by Sir Thomas Heath. Definitive edition of one of very greatest classics of Western world. Complete translation of Heiberg text, plus spurious Book XIV. 150 page introduction on Greek, Medieval mathematics, Euclid, texts, commentators, etc. Elaborate critical apparatus parallels text, analyzing each definition, postulate, proposition, covering textual matters, refutations, supports, extrapolations, etc. This is the full Euclid. Unabridged reproduction of Cambridge U. 2nd edition. 3 volumes. 995 figures. 1426pp. 5⅜ x 8. S88, 89, 90, 3 volume set, paperbound **$6.00**

AN INTRODUCTION TO GEOMETRY OF N DIMENSIONS, D. M. Y. Sommerville. Presupposes no previous knowledge of field. Only book in English devoted exclusively to higher dimensional geometry. Discusses fundamental ideas of incidence, parallelism, perpendicularity, angles between linear space, enumerative geometry, analytical geometry from projective and metric views, polytopes, elementary ideas in analysis situs, content of hyperspacial figures. 60 diagrams. 196pp. 5⅜ x 8. S494 Paperbound **$1.50**

ELEMENTS OF NON-EUCLIDEAN GEOMETRY, D. M. Y. Sommerville. Unique in proceeding step-by-step. Requires only good knowledge of high-school geometry and algebra, to grasp elementary hyperbolic, elliptic, analytic non-Euclidean Geometries; space curvature and its implications; radical axes; homopethic centres and systems of circles; parataxy and parallelism; Gauss' proof of defect area theorem; much more, with exceptional clarity. 126 problems at chapter ends. 133 figures. xvi + 274pp. 5⅜ x 8. S460 Paperbound **$1.50**

THE FOUNDATIONS OF EUCLIDEAN GEOMETRY, H. G. Forder. First connected, rigorous account in light of modern analysis, establishing propositions without recourse to empiricism, without multiplying hypotheses. Based on tools of 19th and 20th century mathematicians, who made it possible to remedy gaps and complexities, recognize problems not earlier discerned. Begins with important relationship of number systems in geometrical figures. Considers classes, relations, linear order, natural numbers, axioms for magnitudes, groups, quasi-fields, fields, non-Archimedian systems, the axiom system (at length), particular axioms (two chapters on the Parallel Axioms), constructions, congruence, similarity, etc. Lists: axioms employed, constructions, symbols in frequent use. 295pp. 5⅜ x 8. S481 Paperbound **$2.00**

CALCULUS, FUNCTION THEORY (REAL AND COMPLEX), FOURIER THEORY

FIVE VOLUME "THEORY OF FUNCTIONS" SET BY KONRAD KNOPP. Provides complete, readily followed account of theory of functions. Proofs given concisely, yet without sacrifice of completeness or rigor. These volumes used as texts by such universities as M.I.T., Chicago, N.Y. City College, many others. "Excellent introduction . . . remarkably readable, concise, clear, rigorous," J. of the American Statistical Association.

ELEMENTS OF THE THEORY OF FUNCTIONS, Konrad Knopp. Provides background for further volumes in this set, or texts on similar level. Partial contents: Foundations, system of complex numbers and Gaussian plane of numbers, Riemann sphere of numbers, mapping by linear functions, normal forms, the logarithm, cyclometric functions, binomial series. "Not only for the young student, but also for the student who knows all about what is in it," Mathematical Journal. 140pp. 5⅜ x 8. S154 Paperbound **$1.35**

THEORY OF FUNCTIONS, PART I, Konrad Knopp. With volume II, provides coverage of basic concepts and theorems. Partial contents: numbers and points, functions of a complex variable, integral of a continuous function, Cauchy's intergral theorem, Cauchy's integral formulae, series with variable terms, expansion and analytic function in a power series, analytic continuation and complete definition of analytic functions, Laurent expansion, types of singularities. vii + 146pp. 5⅜ x 8. S156 Paperbound **$1.35**

THEORY OF FUNCTIONS, PART II, Konrad Knopp. Application and further development of general theory, special topics. Single valued functions, entire, Weierstrass. Meromorphic functions: Mittag-Leffler. Periodic functions. Multiple valued functions. Riemann surfaces. Algebraic functions. Analytical configurations, Riemann surface. x + 150pp. 5⅜ x 8. S157 Paperbound **$1.35**

PROBLEM BOOK IN THE THEORY OF FUNCTIONS, VOLUME I, Konrad Knopp. Problems in elementary theory, for use with Knopp's "Theory of Functions," or any other text. Arranged according to increasing difficulty. Fundamental concepts, sequences of numbers and infinite series, complex variable, integral theorems, development in series, conformal mapping. Answers. viii + 126pp. 5⅜ x 8. S 158 **Paperbound $1.35**

PROBLEM BOOK IN THE THEORY OF FUNCTIONS, VOLUME II, Konrad Knopp. Advanced theory of functions, to be used with Knopp's "Theory of Functions," or comparable text. Singularities, entire and meromorphic functions, periodic, analytic, continuation, multiple-valued functions, Riemann surfaces, conformal mapping. Includes section of elementary problems. "The difficult task of selecting . . . problems just within the reach of the beginner is here masterfully accomplished," AM. MATH. SOC. Answers. 138pp. 5⅜ x 8. S159 Paperbound **$1.35**

ADVANCED CALCULUS, E. B. Wilson. Still recognized as one of most comprehensive, useful texts. Immense amount of well-represented, fundamental material, including chapters on vector functions, ordinary differential equations, special functions, calculus of variations, etc., which are excellent introductions to these areas. Requires only one year of calculus. Over 1300 exercises cover both pure math and applications to engineering and physical problems. Ideal reference, refresher. 54 page introductory review. ix + 566pp. 5⅜ x 8. S504 Paperbound **$2.45**

LECTURES ON THE THEORY OF ELLIPTIC FUNCTIONS, H. Hancock. Reissue of only book in English with so extensive a coverage, especially of Abel, Jacobi, Legendre, Weierstrass, Hermite, Liouville, and Riemann. Unusual fullness of treatment, plus applications as well as theory in discussing universe of elliptic integrals, originating in works of Abel and Jacobi. Use is made of Riemann to provide most general theory. 40-page table of formulas. 76 figures. xxiii + 498pp. 5⅜ x 8. S483 Paperbound **$2.55**

THEORY OF FUNCTIONALS AND OF INTEGRAL AND INTEGRO-DIFFERENTIAL EQUATIONS, Vito Volterra. Unabridged republication of only English translation. General theory of functions depending on continuous set of values of another function. Based on author's concept of transition from finite number of variables to a continually infinite number. Includes much material on calculus of variations. Begins with fundamentals, examines generalization of analytic functions, functional derivative equations, applications, other directions of theory, etc. New introduction by G. C. Evans. Biography, criticism of Volterra's work by E. Whittaker. xxxx + 226pp. 5⅜ x 8. S502 Paperbound **$1.75**

AN INTRODUCTION TO FOURIER METHODS AND THE LAPLACE TRANSFORMATION, Philip Franklin. Concentrates on essentials, gives broad view, suitable for most applications. Requires only knowledge of calculus. Covers complex qualities with methods of computing elementary functions for complex values of argument and finding approximations by charts; Fourier series; harmonic anaylsis; much more. Methods are related to physical problems of heat flow, vibrations, electrical transmission, electromagnetic radiation, etc. 828 problems, answers. Formerly entitled "Fourier Methods." x + 289pp. 5⅜ x 8. S452 Paperbound **$1.75**

THE ANALYTICAL THEORY OF HEAT, Joseph Fourier. This book, which revolutionized mathematical physics, has been used by generations of mathematicians and physicists interested in heat or application of Fourier integral. Covers cause and reflection of rays of heat, radiant heating, heating of closed spaces, use of trigonometric series in theory of heat, Fourier integral, etc. Translated by Alexander Freeman. 20 figures. xxii + 466pp. 5⅜ x 8. S93 Paperbound **$2.00**

ELLIPTIC INTEGRALS, H. Hancock. Invaluable in work involving differential equations with cubics, quatrics under root sign, where elementary calculus methods are inadequate. Practical solutions to problems in mathematics, engineering, physics; differential equations requiring integration of Lamé's, Briot's, or Bouquet's equations; determination of arc of ellipse, hyperbola, lemniscate; solutions of problems in elastics; motion of a projectile under resistance varying as the cube of the velocity; pendulums; more. Exposition in accordance with Legendre-Jacobi theory. Rigorous discussion of Legendre transformations. 20 figures. 5 place table. 104pp. 5⅜ x 8. S484 Paperbound **$1.25**

THE TAYLOR SERIES, AN INTRODUCTION TO THE THEORY OF FUNCTIONS OF A COMPLEX VARIABLE, P. Dienes. Uses Taylor series to approach theory of functions, using ordinary calculus only, except in last 2 chapters. Starts with introduction to real variable and complex algebra, derives properties of infinite series, complex differentiation, integration, etc. Covers biuniform mapping, overconvergence and gap theorems, Taylor series on its circle of convergence, etc. Unabridged corrected reissue of first edition. 186 examples, many fully worked out. 67 figures. xii + 555pp. 5⅜ x 8. S391 Paperbound **$2.75**

LINEAR INTEGRAL EQUATIONS, W. V. Lovitt. Systematic survey of general theory, with some application to differential equations, calculus of variations, problems of math, physics. Includes: integral equation of 2nd kind by successive substitutions; Fredholm's equation as ratio of 2 integral series in lambda, applications of the Fredholm theory, Hilbert-Schmidt theory of symmetric kernels, application, etc. Neumann, Dirichlet, vibratory problems. ix + 253pp. 5⅜ x 8. S175 Clothbound **$3.50** S176 Paperbound **$1.60**

DOVER SCIENCE BOOKS

DICTIONARY OF CONFORMAL REPRESENTATIONS, H. Kober. Developed by British Admiralty to solve Laplace's equation in 2 dimensions. Scores of geometrical forms and transformations for electrical engineers, Joukowski aerofoil for aerodynamics, Schwartz-Christoffel transformations for hydro-dynamics, transcendental functions. Contents classified according to analytical functions describing transformations with corresponding regions. Glossary. Topological index. 447 diagrams. 6⅛ x 9¼.   .S160 Paperbound **$2.00**

ELEMENTS OF THE THEORY OF REAL FUNCTIONS, J. E. Littlewood. Based on lectures at Trinity College, Cambridge, this book has proved extremely successful in introducing graduate students to modern theory of functions. Offers full and concise coverage of classes and cardinal numbers, well ordered series, other types of series, and elements of the theory of sets of points. 3rd revised edition. vii + 71pp. 5⅜ x 8.   S171 Clothbound **$2.85**
                                                                                        S172 Paperbound **$1.25**

INFINITE SEQUENCES AND SERIES, Konrad Knopp. 1st publication in any language. Excellent introduction to 2 topics of modern mathematics, designed to give student background to penetrate further alone. Sequences and sets, real and complex numbers, etc. Functions of a real and complex variable. Sequences and series. Infinite series. Convergent power series. Expansion of elementary functions. Numerical evaluation of series. v + 186pp. 5⅜ x 8.
                                                                                        S152 Clothbound **$3.50**
                                                                                        S153 Paperbound **$1.75**

THE THEORY AND FUNCTIONS OF A REAL VARIABLE AND THE THEORY OF FOURIER'S SERIES, E. W .Hobson. One of the best introductions to set theory and various aspects of functions and Fourier's series. Requires only a good background in calculus. Exhaustive coverage of: metric and descriptive properties of sets of points; transfinite numbers and order types; functions of a real variable; the Riemann and Lebesgue integrals; sequences and series of numbers; power-series; functions representable by series sequences of continuous functions; trigonometrical series; representation of functions by Fourier's series; and much more. "The best possible guide," Nature. Vol. I: 88 detailed examples, 10 figures. Index. xv + 736pp. Vol. II: 117 detailed examples, 13 figures. x + 780pp. 6⅛ x 9¼.
                                                                                        Vol. I: S387 Paperbound **$3.00**
                                                                                        Vol. II: S388 Paperbound **$3.00**

ALMOST PERIODIC FUNCTIONS, A. S. Besicovitch. Unique and important summary by a well known mathematician covers in detail the two stages of development in Bohr's theory of almost periodic functions: (1) as a generalization of pure periodicity, with results and proofs; (2) the work done by Stepanof, Wiener, Weyl, and Bohr in generalizing the theory. xi + 180pp. 5⅜ x 8.   S18 Paperbound **$1.75**

INTRODUCTION TO THE THEORY OF FOURIER'S SERIES AND INTEGRALS, H. S. Carslaw. 3rd revised edition, an outgrowth of author's courses at Cambridge. Historical introduction, rational, irrational numbers, infinite sequences and series, functions of a single variable, definite integral, Fourier series, and similar topics. Appendices discuss practical harmonic analysis, periodogram analysis, Lebesgue's theory. 84 examples. xiii + 368pp. 5⅜ x 8.
                                                                                        S48 Paperbound **$2.00**

SYMBOLIC LOGIC

THE ELEMENTS OF MATHEMATICAL LOGIC, Paul Rosenbloom. First publication in any language. For mathematically mature readers with no training in symbolic logic. Development of lectures given at Lund Univ., Sweden, 1948. Partial contents: Logic of classes, fundamental theorems, Boolean algebra, logic of propositions, of propositional functions, expressive languages, combinatory logics, development of math within an object language, paradoxes, theorems of Post, Goedel, Church, and similar topics. iv + 214pp. 5⅜ x 8.
                                                                                        S227 Paperbound **$1.45**

INTRODUCTION TO SYMBOLIC LOGIC AND ITS APPLICATION, R. Carnap. Clear, comprehensive, rigorous, by perhaps greatest living master. Symbolic languages analyzed, one constructed. Applications to math (axiom systems for set theory, real, natural numbers), topology (Dedekind, Cantor continuity explanations), physics (general analysis of determination, causality, space-time topology), biology (axiom system for basic concepts). "A masterpiece," Zentralblatt für Mathematik und Ihre Grenzgebiete. Over 300 exercises. 5 figures. xvi + 241pp. 5⅜ x 8.   S453 Paperbound **$1.85**

AN INTRODUCTION TO SYMBOLIC LOGIC, Susanne K. Langer. Probably clearest book for the philosopher, scientist, layman—no special knowledge of math required. Starts with simplest symbols, goes on to give remarkable grasp of Boole-Schroeder, Russell-Whitehead systems, clearly, quickly. Partial Contents: Forms, Generalization, Classes, Deductive System of Classes, Algebra of Logic, Assumptions of Principia Mathematica, Logistics, Proofs of Theorems, etc. "Clearest . . . simplest introduction . . . the intelligent non-mathematician should have no difficulty," MATHEMATICS GAZETTE. Revised, expanded 2nd edition. Truth-value tables. 368pp. 5⅜   8.   S164 Paperbound **$1.75**

TRIGONOMETRICAL SERIES, Antoni Zygmund. On modern advanced level. Contains carefully organized analyses of trigonometric, orthogonal, Fourier systems of functions, with clear adequate descriptions of summability of Fourier series, proximation theory, conjugate series, convergence, divergence of Fourier series. Especially valuable for Russian, Eastern European coverage. 329pp. 5⅜ x 8. S290 Paperbound **$1.50**

THE LAWS OF THOUGHT, George Boole. This book founded symbolic logic some 100 years ago. It is the 1st significant attempt to apply logic to all aspects of human endeavour. Partial contents: derivation of laws, signs and laws, interpretations, eliminations, conditions of a perfect method, analysis, Aristotelian logic, probability, and similar topics. xvii + 424pp. 5⅜ x 8. S28 Paperbound **$2.00**

SYMBOLIC LOGIC, C. I. Lewis, C. H. Langford. 2nd revised edition of probably most cited book in symbolic logic. Wide coverage of entire field; one of fullest treatments of paradoxes; plus much material not available elsewhere. Basic to volume is distinction between logic of extensions and intensions. Considerable emphasis on converse substitution, while matrix system presents supposition of variety of non-Aristotelian logics. Especially valuable sections on strict limitations, existence theorems. Partial contents: Boole-Schroeder algebra; truth value systems, the matrix method; implication and deductibility; general theory of propositions; etc. "Most valuable," Times, London. 506pp. 5⅜ x 8. S170 Paperbound **$2.00**

GROUP THEORY AND LINEAR ALGEBRA, SETS, ETC.

LECTURES ON THE ICOSAHEDRON AND THE SOLUTION OF EQUATIONS OF THE FIFTH DEGREE, Felix Klein. Solution of quintics in terms of rotations of regular icosahedron around its axes of symmetry. A classic, indispensable source for those interested in higher algebra, geometry, crystallography. Considerable explanatory material included. 230 footnotes, mostly bibliography. "Classical monograph . . . detailed, readable book," Math. Gazette. 2nd edition. xvi + 289pp. 5⅜ x 8. S314 Paperbound **$1.85**

INTRODUCTION TO THE THEORY OF GROUPS OF FINITE ORDER, R. Carmichael. Examines fundamental theorems and their applications. Beginning with sets, systems, permutations, etc., progresses in easy stages through important types of groups: Abelian, prime power, permutation, etc. Except 1 chapter where matrices are desirable, no higher math is needed. 783 exercises, problems. xvi + 447pp. 5⅜ x 8. S299 Clothbound **$3.95**
S300 Paperbound **$2.00**

THEORY OF GROUPS OF FINITE ORDER, W. Burnside. First published some 40 years ago, still one of clearest introductions. Partial contents: permutations, groups independent of representation, composition series of a group, isomorphism of a group with itself, Abelian groups, prime power groups, permutation groups, invariants of groups of linear substitution, graphical representation, etc. "Clear and detailed discussion . . . numerous problems which are instructive," Design News. xxiv + 512pp. 5⅜ x 8. S38 Paperbound **$2.45**

COMPUTATIONAL METHODS OF LINEAR ALGEBRA, V. N. Faddeeva, translated by C. D. Benster. 1st English translation of unique, valuable work, only one in English presenting systematic exposition of most important methods of linear algebra—classical, contemporary. Details of deriving numerical solutions of problems in mathematical physics. Theory and practice. Includes survey of necessary background, most important methods of solution, for exact, iterative groups. One of most valuable features is 23 tables, triple checked for accuracy, unavailable elsewhere. Translator's note. x + 252pp. 5⅜ x 8. S424 Paperbound **$1.95**

THE CONTINUUM AND OTHER TYPES OF SERIAL ORDER, E. V. Huntington. This famous book gives a systematic elementary account of the modern theory of the continuum as a type of serial order. Based on the Cantor-Dedekind ordinal theory, which requires no technical knowledge of higher mathematics, it offers an easily followed analysis of ordered classes, discrete and dense series, continuous series, Cantor's transfinite numbers. "Admirable introduction to the rigorous theory of the continuum . . . reading easy," Science Progress. 2nd edition. viii + 82pp. 5⅜ x 8. S129 Clothbound **$2.75**
S130 Paperbound **$1.00**

THEORY OF SETS, E. Kamke. Clearest, amplest introduction in English, well suited for independent study. Subdivisions of main theory, such as theory of sets of points, are discussed, but emphasis is on general theory. Partial contents: rudiments of set theory, arbitrary sets, their cardinal numbers, ordered sets, their order types, well-ordered sets, their cardinal numbers. vii + 144pp. 5⅜ x 8. S141 Paperbound **$1.35**

CONTRIBUTIONS TO THE FOUNDING OF THE THEORY OF TRANSFINITE NUMBERS, Georg Cantor. These papers founded a new branch of mathematics. The famous articles of 1895-7 are translated, with an 82-page introduction by P. E. B. Jourdain dealing with Cantor, the background of his discoveries, their results, future possibilities. ix + 211pp. 5⅜ x 8. S45 Paperbound **$1.25**

14

DOVER SCIENCE BOOKS

NUMERICAL AND GRAPHICAL METHODS, TABLES

JACOBIAN ELLIPTIC FUNCTION TABLES, L. M. Milne-Thomson. Easy-to-follow, practical, not only useful numerical tables, but complete elementary sketch of application of elliptic functions. Covers description of principle properties; complete elliptic integrals; Fourier series, expansions; periods, zeros, poles, residues, formulas for special values of argument; cubic, quartic polynomials; pendulum problem; etc. Tables, graphs form body of book: Graph, 5 figure table of elliptic function sn (u m); cn (u m); dn (u m). 8 figure table of complete elliptic integrals K, K', E, E', nome q. 7 figure table of Jacobian zeta-function Z(u). 3 figures. xi + 123pp. 5⅜ x 8. S194 Paperbound **$1.35**

TABLES OF FUNCTIONS WITH FORMULAE AND CURVES, E. Jahnke, F. Emde. Most comprehensive 1-volume English text collection of tables, formulae, curves of transcendent functions. 4th corrected edition, new 76-page section giving tables, formulae for elementary functions not in other English editions. Partial contents: sine, cosine, logarithmic integral; error integral; elliptic integrals; theta functions; Legendre, Bessel, Riemann, Mathieu, hypergeometric functions; etc. "Out-of-the-way functions for which we know no other source." Scientific Computing Service, Ltd. 212 figures. 400pp. 5⅜ x 8⅜. S133 Paperbound **$2.00**

MATHEMATICAL TABLES, H. B. Dwight. Covers in one volume almost every function of importance in applied mathematics, engineering, physical sciences. Three extremely fine tables of the three trig functions, inverses, to 1000th of radian; natural, common logs; squares, cubes; hyperbolic functions, inverses; $(a^2 + b^2)$ exp: ½a; complete elliptical integrals of 1st, 2nd kind; sine, cosine integrals; exponential integrals; Ei(x) and Ei($-$x); binomial coefficients; factorials to 250; surface zonal harmonics, first derivatives; Bernoulli, Euler numbers, their logs to base of 10; Gamma function; normal probability integral; over 60pp. Bessel functions; Riemann zeta function. Each table with formulae generally used, sources of more extensive tables, interpolation data, etc. Over half have columns of differences, to facilitate interpolation. viii + 231pp. 5⅜ x 8. S445 Paperbound **$1.75**

PRACTICAL ANALYSIS, GRAPHICAL AND NUMERICAL METHODS, F. A. Willers. Immensely practical hand-book for engineers. How to interpolate, use various methods of numerical differentiation and integration, determine roots of a single algebraic equation, system of linear equations, use empirical formulas, integrate differential equations, etc. Hundreds of short-cuts for arriving at numerical solutions. Special section on American calculating machines, by T. W. Simpson. Translation by R. T. Beyer. 132 illustrations. 422pp. 5⅜ x 8. S273 Paperbound **$2.00**

NUMERICAL SOLUTIONS OF DIFFERENTIAL EQUATIONS, H. Levy, E. A. Baggott. Comprehensive collection of methods for solving ordinary differential equations of first and higher order. 2 requirements: practical, easy to grasp; more rapid than school methods. Partial contents: graphical integration of differential equations, graphical methods for detailed solution. Numerical solution. Simultaneous equations and equations of 2nd and higher orders. "Should be in the hands of all in research and applied mathematics, teaching," Nature. 21 figures. viii + 238pp. 5⅜ x 8. S168 Paperbound **$1.75**

NUMERICAL INTEGRATION OF DIFFERENTIAL EQUATIONS, Bennet, Milne, Bateman. Unabridged republication of original prepared for National Research Council. New methods of integration by 3 leading mathematicians: "The Interpolational Polynomial," "Successive Approximation," A. A. Bennett, "Step-by-step Methods of Integration," W. W. Milne. "Methods for Partial Differential Equations," H. Bateman. Methods for partial differential equations, solution of differential equations to non-integral values of a parameter will interest mathematicians, physicists. 288 footnotes, mostly bibliographical. 235 item classified bibliography. 108pp. 5⅜ x 8. S305 Paperbound **$1.35**

Write for free catalogs!

Indicate your field of interest. Dover publishes books on physics, earth sciences, mathematics, engineering, chemistry, astronomy, anthropology, biology, psychology, philosophy, religion, history, literature, mathematical recreations, languages, crafts, art, graphic arts, etc.

Write to Dept. catr
Dover Publications, Inc.
Science A
180 Varick St., N. Y. 14, N. Y.

15